T0296576

CAMBRIDGE MONOGRAPHS ON PHYSICS

GENERAL EDITORS

N. FEATHER, F.R.S.
Professor of Natural Philosophy in the University of Edinburgh

D. SHOENBERG, PH.D.
Fellow of Gonville & Caius College, Cambridge

EXCITED STATES OF NUCLEI

EXCITED STATES OF NUCLEI

BY

S. DEVONS, Ph.D.
Fellow of Trinity College, Cambridge

CAMBRIDGE

AT THE UNIVERSITY PRESS

1949

CAMBRIDGE UNIVERSITY PRESS
Cambridge, New York, Melbourne, Madrid, Cape Town,
Singapore, São Paulo, Delhi, Mexico City

Cambridge University Press
The Edinburgh Building, Cambridge CB2 8RU, UK

Published in the United States of America by Cambridge University Press, New York

www.cambridge.org
Information on this title: www.cambridge.org/9781107621411

First published 1949
First paperback edition 2013

A catalogue record for this publication is available from the British Library

ISBN 978-1-107-62141-1 Paperback

GENERAL PREFACE

The Cambridge Physical Tracts, out of which this series of Monographs has developed, were planned and originally published in a period when book production was a fairly rapid process. Unfortunately, that is no longer so, and to meet the new situation a change of title and a slight change of emphasis have been decided on. The major aim of the series will still be the presentation of the results of recent research, but individual volumes will be somewhat more substantial, and more comprehensive in scope, than were the volumes of the older series. This will be true, in many cases, of new editions of the Tracts, as these are republished in the expanded series, and it will be true in most cases of the Monographs which have been written since the War or are still to be written.

The aim will be that the series as a whole shall remain representative of the entire field of pure physics, but it will occasion no surprise if, during the next few years, the subject of nuclear physics claims a large share of attention. Only in this way can justice be done to the enormous advances in this field of research over the War years.

N. F.
D. S.

CONTENTS

CHAPTER 5
Nuclear Spectra and their Interpretation

18. Scope of theoretical interpretation, p. 105. **19.** Symmetrical force theory, p. 106. **19·1.** Wigner's theory of 'super-multiplet' structure, p. 108. **20.** Nuclear models, p. 117. **20·1.** Independent-particle model, p. 118. **20·2.** Hartree approximation, p. 120. **20·3.** α-Particle model, p. 124. **20·4.** Rigid-body model, p. 128. **20·5.** Liquid-drop model, p. 132. **21.** Statistical properties of models, p. 133. **21·1.** Independent-particle model, p. 134. **21·2.** Liquid-drop model, p. 138. **22.** Some experimental regularities, p. 141. **22·1.** Integral relations between excitation energies, p. 141. **22·2.** Constancy of level spacing, p. 142. **22·3.** Density of resonance levels, p. 143. **22·4.** Occurrence of isomeric states, p. 144.

REFERENCES   *page* 146

SUBJECT INDEX   *page* 151

AUTHOR'S PREFACE

In attempting to give a short account of the study of nuclear energy levels I have tried to steer an uneasy course between the dangers, on the one hand, of presenting only the more established experimental material and uncontroversial theoretical interpretations which would mean avoiding much that is topical, and, on the other, of being so preoccupied with the most recent work as to make it probable that the account would be largely obsolete by the time it appeared in print. It is easy enough to recognize these dangers, but to avoid them, in a subject in which new experimental material is being obtained at such a prodigious rate (some four hundred papers dealing with this subject have been published in the past twelve months), must be a matter of chance as well of skill. I hope that I have not relied on the former unduly.

In the main the book is written from the experimental standpoint, i.e. both the technique and scope, as well as the results, of experimental work are discussed, whereas only the results of theoretical work are, in general, indicated. It has not been possible to include all the experimental work which would be required to give a comprehensive account of this subject. Typical experimental material has therefore been chosen to serve as illustrations of both the techniques and the empirical information available. Likewise the scope of the theoretical interpretations has not been dealt with in detail, since quite a cursory comparison of theory and experiment will reveal the limitations of the former and the incompleteness of the latter.

My thanks are due to Professor O. R. Frisch and Professor N. Feather for helpful comments on some parts of the book, and to the appropriate bodies for permission to reproduce diagrams from the *Proceedings of the Royal Society*, the *Physical Review*, and the *Proceedings of the Physical Society*.

S. D.

MARCH 1949

CHAPTER I

INTRODUCTION

1·1. Properties of energy levels

Any system of particles occupying a limited volume of space and in a state of dynamical equilibrium, must, according to the principles of quantum mechanics, have a total energy equal to one of a discrete set of values, the 'energy levels' of the system. Each of these energy levels corresponds either to a particular equilibrium (i.e. 'stationary') state of the system, or if the energy levels are degenerate, to a set of stationary states. These energies are eigen-values of the Hamiltonian operator representing the system of particles, and the stationary states are represented by the corresponding eigen-functions of the Hamiltonian. The levels are degenerate if a number of eigen-functions correspond to the same eigen-value. If the system is not completely in equilibrium but can be characterized by small, finite probabilities of making transitions to other states, not spatially restricted, then the energy levels of the system will not be exactly defined. Associated with the transition probabilities there will be finite level widths, Γ, such that if the system has a mean lifetime τ in the approximately stable state, then $\Gamma \sim \hbar/\tau$. Provided the transition probabilities, and therefore also the level widths, are sufficiently small the energy levels may still be regarded as discrete and the properties of the system will be determined, apart from possible degeneracy of the energy levels, by a knowledge of the particular energy level in which the system is formed. The wave functions describing any one of these approximately stationary states will be similar to those applicable to the exactly stationary states, i.e. the wave function will be an eigen-function of some Hamiltonian, $\mathcal{H}_0$, which represents the system accurately except for the small probability of transitions. These latter can be included by associating with each eigen-function of $\mathcal{H}_0$ additional quantities representing the probabilities of the various energetically possible transitions. $\mathcal{H}_0$ also determines the lowest, completely stationary state of the system so that we can regard the energy-level spectrum of a system as discrete over the range of energies for which the state of the system can be described by a wave function corresponding to a single eigen-value of $\mathcal{H}_0$, the Hamiltonian used to determine the ground state.

If the widths of the actual energy levels are comparable with the level separation, so that appreciable overlapping of levels occurs, then it is no longer possible to describe the state of the system in terms of the eigen-functions belonging to a single eigen-value of $\mathcal{H}_0$; the description in terms of eigen-functions of $\mathcal{H}_0$ involves a linear superposition of eigen-functions belonging to several different eigen-values of $\mathcal{H}_0$. This superposition will involve definite phase relations between the component eigen-functions, and these phase differences, and hence also the various transition probabilities and other properties of the state, will be determined by the particular manner in which the system is formed. It is no longer possible to characterize the system by a description limited to the position and properties of the energy levels, and therefore the value of these concepts is considerably reduced.

In the present account of nuclear energy levels we shall deal principally with nuclear states which belong to that region of the energy-level spectrum in which the discrete nature of the level spectrum is still strongly in evidence, i.e. we shall be dealing with approximately stationary states with well-defined properties and which can be described in principle without any reference to a particular mode of formation. There is, of course, no sharp line of demarcation between those parts of the level spectrum in which states are discrete and those in which overlapping occurs. The extremes are easily recognized. For states of low energy for which the only energetically possible transitions are by electromagnetic radiation, the widths of the levels are extremely small and over-lapping of levels can be completely ignored. For very high energies, the time during which all the constituent nucleons are together in a region of nuclear dimensions may be so small (i.e. only of the same order as the characteristic nuclear time, 10^{-21} sec.), that the general concept of a stationary state of the whole nucleus is irrelevant. For a limited range of intermediate energies, effectively discrete levels occur if the dissociation of one or more nucleons from the nuclear system is a sufficiently slow (i.e. improbable) process. Such levels will be referred to as *virtual* levels of the nucleus if the dissociation from the nucleus of one or more nucleons is energetically possible, and as *bound* levels if the only transition possible is by electromagnetic radiation.

1·2. Phenomenological and theoretical viewpoints

The study of nuclear energy levels can be regarded from two viewpoints which for brevity we call the *phenomenological* and *theoretical*. From the first of these viewpoints the essential problem is the *experimental* determination of the energies and transition probabilities of individual nuclear energy levels or the deduction of these essential quantities from particular experimental data. Once established, such a set of parameters can be used to interpret, correlate and predict a range of nuclear phenomena, particularly in the field of nuclear disintegrations and scattering. This approach to the problem is of great practical importance. Its interest from the theoretical point of view is twofold. In the first place it indicates the validity of theoretical principles well established in other branches of physics, albeit for a limited interpretation of nuclear phenomena. Secondly, it reduces a large amount of experimental material to a minimum number of empirically determined quantities which form a suitable basis for comparison with theory. The 'theory' is now concerned not with predictions about nuclear processes *from* a knowledge of nuclear levels but rather with the *prediction* of the level positions and properties themselves, or more generally with the problem of the interaction of nucleons and the structure of nuclei. Whether or not the normal quantum-mechanical methods for dealing with systems of many particles, and in particular the construction of a Hamiltonian which is a function of the co-ordinates of nucleons and some assumed laws of interaction between them, are in fact justifiable for interpretation of nuclear phenomena is best judged by the results of such attempts. In the discussion that follows we shall assume that this procedure is justifiable (or at least that no better procedure is known), and that the shortcomings of the theoretical treatments are attributable to imprecise knowledge of the interactions between nucleons and the difficulties inherent in the application of the usual quantum-mechanical methods to complicated nuclear systems.

The phenomenological approach has been developed extensively and successfully, especially since Bohr first indicated the significance of the *many-body* nature of nuclear processes. The formal development of the *Dispersion Formula* for nuclear processes and the application of the principles of statistical mechanics have provided a clear indication of the general validity and scope of the phenomenological method and confirmed the view that the nucleus must be treated as a complex, many-body system. But from the theoretical point of view the emphasis on the many-body aspect, justified by the success of the developments it has engendered, only serves to stress the difficulties that are involved in any attempts to formulate a theory of nuclear structure comparable in scope, for example, with that of atomic or molecular structure.

1·3. Spectrum of levels

The dominant feature governing nuclear structure and the interaction of nucleons with nuclei is the short range of the forces between nucleons. As a result of this property of the forces, the nucleus forms an essentially *closed* structure, i.e. an impinging nucleon cannot pass through the nucleus without interacting strongly with many of the constituent nucleons; in fact, in general it must be regarded as interacting with the nucleus as a whole. If it penetrates to the surface of the nucleus it will, normally, form a 'compound' nucleus which will exist for a time large in comparison with that taken for the nucleon to traverse a distance of the order of nuclear dimensions;* indeed, the incident nucleon, or any other, will not escape from the system until a chance redistribution concentrates sufficient energy on a single nucleon (or group) sufficient for it to dissociate from the remainder. Such a system, in an approximately stationary state, will have a well-defined energy, and if the energy is not too large the energy levels will be practically discrete, as is, in fact, found to be the case experimentally. Thus the short-range nature of nuclear forces and the many-body nature of the nucleus can explain the occurrence of discrete virtual levels of nuclei. It also suggests the type of energy-level spectrum to be expected. A fixed potential field of the short-range type, e.g. a deep

* For a characteristic kinetic energy of the nucleon of 5 MeV. and a medium-size nucleus this time is of the order 10^{-21} sec.

rectangular potential well, gives rise to a set of energy levels quite different from those associated with, for example, the long-range Coulomb forces of an atom (fig. 1 *a*, *b*). Characteristic of the short-range forces is the fact that the spacing of the levels does not approach zero asymptotically as the energy approaches zero (as in the case of the Coulomb potential). Thus for a short-range force there is a finite number of bound levels. Of course the actual energy levels of the nucleus cannot be represented by the levels of a single particle in a fixed potential well; in fact, the decrease in level spacing with increasing energy will be much more rapid for a system of

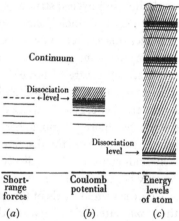

Fig. 1. Energy levels for short- and long-range forces.

particles than for a single particle, but the essential feature, a finite number of bound levels, remains.

When the excitation energy of the nucleus exceeds that required to dissociate from it a nucleon (or group of nucleons) the energy levels may still be very narrow and the spectrum of levels not essentially different from the spectrum at excitation energies just less than the dissociation energy. For a nucleus, then, in marked contrast with the energy levels of optical spectra, one would not expect any change in character of the level spectrum above and below the dissociation limit.

1·4. Level spectrum and structure

The energy-level spectrum of a system is characteristic not only of the forces between the constituents but of the structure of the system. Thus for an atomic system with a single predominant centre of force and mass it is possible to consider, with success, that each electron moves in a fixed static field, and consequently can be characterized by its individual energy level. The atom can be excited so that an electron in, say, the K shell, is moved into any one of the vacant outer orbits, i.e. a lightly bound level. In this way the energy-level spectrum of the whole atom can be regarded as a repetition, with varying energy intervals, of the basic pattern of the optical level spectrum, this repetition being characteristic of the *shell structure* of the atom (fig. 1c). Similar grouping of levels, corresponding to regularities in physical structure are exhibited by molecular spectra. In the case of nuclei, however, no such well-defined structure has been discovered, nor perhaps should it be expected for a system comprising many similar strongly interacting particles with no single strong centre of force. The nuclear level spectrum would be expected to show a progressive reduction in level spacing as the excitation energy is decreased, but no marked grouping into characteristic energy regions has been observed. Incidentally, it should be noticed that the discrete virtual levels of an atom (i.e. levels with sufficient energy to dissociate an electron) are well defined in energy because of the *weak* interaction between individual electrons, whereas virtual levels of the nucleus owe their discrete nature to the *strong* interaction between the many constituent nucleons. This difference is illustrated by the difference in physical properties of states of the two types. An atom excited by ejection of a K electron will normally lose energy by electromagnetic radiation; a nucleus with excitation energy greatly exceeding the dissociation limit will normally eject one or more nucleons.

1·5. Scope of investigations

Apart from the theoretically significant differences between the spectra of nuclei and of atoms and molecules, technical differences in available experimental procedures lead to a different emphasis in studying spectra of the two types. The most outstanding feature of the study of atomic spectra is the great precision of the measurement, particularly in the optical region, and the relative simplicity of the methods of exciting complete atomic spectra. Nuclear spectra cannot be excited so easily, nor can the radiation be measured with comparable precision or resolution. More emphasis is therefore placed on a study of the properties of individual nuclear levels, particularly in the case of virtual levels where nucleon dissociation as well as electromagnetic radiation can occur. The limiting excitation energy for bound levels varies from zero to about 10 MeV. For light nuclei ($Z < 10$) the number of bound levels appears sufficiently small for the determination of their individual excitation energies and properties to be possible. For heavier nuclei the absolute positions of the lower levels, perhaps up to 4 or 5 MeV. excitation energy, may be significant and experimentally ascertainable with a reasonable expenditure of effort. For higher energies, including those of the virtual levels, the density, grouping, transition probabilities and other characteristics of the levels, rather than their exact location, are the properties that form the subject of investigations.

CHAPTER II

EXCITATION OF BOUND STATES

2. EXCITATION IN SCATTERING AND IN NUCLEAR TRANSMUTATIONS[1,2]

Although the transition in the energy-level spectrum of nuclei between bound and virtual excited states is, in most respects, a smooth one, we shall, nevertheless, classify the experimental procedures employed according to the type of nuclear level about which they provide information. In many cases, particularly in the study of nuclear transmutations by bombardment with nucleons, information about both types of level may be obtained more or less simultaneously, but the particular conditions of an experiment must usually be arranged differently if emphasis is placed on information of one type or the other. This classification is particularly convenient where we are concerned with the *scope* of the various experimental methods.

The energy of even the first excited state of most nuclei is some 10^4 eV. or more in excess of the stable* ground state. It is therefore impractical to excite such levels by thermal means, at least with temperatures of a few thousand degrees available in the laboratory.†
Excitation of nuclei by irradiation with beams of high-energy electrons or electromagnetic radiation, analogous to the methods used in the study of atomic spectra, have been successfully employed; but, owing to the strong interaction between such radiations and the extranuclear structure and their relatively weak interactions with the nucleus itself, and also on account of the technical problems of controlling the energy of the incident radiation and detecting the process of excitation (or de-excitation), such methods have not played such a predominant role in the study of

---

* Throughout this book reference to the ground state of a nucleus will be to the state of lowest energy which does not emit electromagnetic radiation. It may either be stable in the usual sense, or β-radioactive.

† Even if in some instances the first excited state happened to be only a few electron volts above the ground state, the 'lifetime' of such an excited nucleus would be so long (small radiation width) that the detection of the feeble radiation would be very difficult.

nuclear spectra as their counterparts in atomic spectra. We shall return to a more detailed discussion of such methods after dealing with the two most widely used procedures for examining bound nuclear states, viz. (a) the excitation by collision with a nucleon, or group of nucleons, in which either a disintegration occurs leaving the residual nucleus in an excited state or the colliding nucleon is scattered inelastically, (b) examination of the β-spectrum and subsequent γ-radiation from β-radioactive nuclei.

For the purpose of studying bound nuclear states, the type of nuclear collision that is most useful can usually be represented by the scheme

$$A(Z, N) + P(p, x) \rightarrow C(Z+p, \ N+x)$$
$$\rightarrow B(Z+p-q, \ N+x-y) + Q(q, y), \qquad (2 \cdot 1)$$

where the symbols in parentheses denote the numbers of protons and neutrons in the respective nuclei. In most cases of interest, where moderate collision energies (W_P) are used,* Q will be either a simple particle or a stable group (α-particle), and the possibility of Q being produced in an excited state will not arise; this possibility will be ignored unless stated specifically to the contrary. The nucleus B, on the other hand, will not in general be produced in its ground state. Denoting by E_P, E_Q the binding energies of the particles P, Q in the compound nucleus C, then $E_{\text{max.}}$, the maximum energy available for excitation of B, will be equal to $E_P - E_Q + W_P$, corresponding to zero energy of emission of particles Q, and it is energetically possible for all levels with excitation energies less than $E_{\text{max.}}$ to be excited. The actual number of nuclei B produced in the individual excited states will depend on the relative probability of the compound nucleus C emitting a particle Q with the corresponding kinetic energy, these different modes of disintegration of C being in mutual competition. Since the compound nucleus C is produced in a state which is unstable with respect to emission of nucleons (the emission of both P and Q at least are possible) its lifetime will normally be very short ($\ll 10^{-16}$ sec.), and the width of the states in which C is formed will be correspondingly large ($\geqslant 1$ eV.). In general, the energy of excitation E' of C will be very

---

* All kinetic energies W refer to measurements made in a co-ordinate frame for which the centre of gravity of the whole system is at rest.

high $(E' = E_P + W_P)$, so that the density of levels will be large, and it is usually the case that the compound nucleus C must be regarded as being formed in a large number of broad overlapping states.

From statistical-mechanical considerations of the relation between the emission of Q from C with energy W_Q and the inverse process, i.e. formation of C by collision of Q with energy W_Q and the appropriate *excited* nucleus B, it can be shown that the probability per unit time, $\Gamma(W_Q)$, of the former process is related to the cross-section for formation in the latter, $\sigma(W_Q)$, by the relation [3]

$$\Gamma(W_Q) = (m\sigma(W_Q) W_Q)/(\pi^2 \hbar^2 \rho_C(E')), \qquad (2\cdot2)$$

where $\rho_C(E')$ is the density of levels of the compound nucleus C at excitation energy E' and m is the reduced mass of Q in the system $B + Q$. The probability Γ is expressed here, as generally, in energy units, i.e. $\Gamma = \hbar/\tau$, where τ is the (partial) lifetime for the particular process.

If the distance between the levels of B, corresponding to the relevant range of energies W_Q, is much smaller than the experimental resolution (determined by the homogeneity of particles Q and the target thickness), then the relative number of particles Q produced in a given *energy interval* is the significant quantity rather than the number corresponding to a particular level of B. Denoting by $\rho_B(E)$ the density of levels of the product nucleus B with energy $E(= W_{\max.} - W_Q)$, then from (2·2) we obtain for the distribution in energy of particles Q from a compound nucleus C with energy E', the expression

$$N(W_Q)\,dW_Q = \alpha(E')\,\sigma(W_Q)\,W_Q(\rho_B(W_{\max.} - W_Q)), \qquad (2\cdot3)$$

where $\alpha(E')$ depends on E' but not on W_Q.

If we define a quantity S by $S_B(E) = \kappa \log \zeta \rho_B(E)$ and T by $(\partial S/\partial E)_T = 1/T(E)$, then if $W_Q \ll W_{\max.}$ we can write

$$S_B(E) = S_B(W_{\max.}) - (\partial S/\partial E)_{W_{\max.}} W_Q$$
$$= S(W_{\max.}) - W_Q/T_B(W_{\max.}); \qquad (2\cdot4)$$

substituting in (2·3) we get

$$N(W_Q)\,dW_Q \sim \alpha(E')\,\sigma(W_Q) \exp\left[-W_Q/\kappa T_B(W_{\max.})\right] dW_Q.$$
$$(2\cdot5)$$

Since $W_{\text{max.}} = E_{\text{max.}}$, the maximum excitation energy of B possible, we can write for the relative probability of B being produced in an excited level of energy E,

$$P(E)\,dE \sim \alpha\sigma(W_{\text{max.}} - E)(W_{\text{max.}} - E)\exp\left[E/\kappa T_B(E_{\text{max.}})\right]dE. \tag{2.6}$$

S and T as defined correspond closely to the entropy and temperature of a system with many degrees of freedom (here the residual excited nucleus B), and this correspondence can be clearly seen if some approximation is made about the dependence of σ on W_Q.*

2·1. Uncharged particles

If $W_Q \gtrsim 2$ MeV. and Q is uncharged, then $\sigma(W_Q)$ will be roughly constant and equal to the geometrical cross-section of the nucleus B. Of course there will be variations from individual level to level of B due, apart from possible other factors, to variations in the relative statistical weights, $(2J + 1)$ and $(2s + 1)(2i + 1)$, of the compound nucleus C (total angular momentum J) and the system $B + Q$ (angular momenta s, i, respectively). However, if the average over many levels is taken these and other variations may be assumed to be unimportant. With this approximation (2·5) indicates that energy distribution of particles Q has a Maxwell distribution with characteristic temperature $T(E_{\text{max.}})$. The process is analogous, as pointed out by Bohr, to the evaporation of molecules from a heated liquid drop. In both cases the temperature T is related to the mean kinetic energy of particles (Q and vapour molecules in the two cases) in equilibrium with B (the drop) necessary to give the latter a mean excitation energy $E_{\text{max.}}$. There are important differences between the two cases, first, due to the discrete nature of the lower levels of B (ignored in the above treatment), and secondly, because the emission of a nucleon Q from the compound nucleus C usually requires a substantial fraction of the total energy of excitation of C. In the liquid drop case the energy required to evaporate a single molecule is insignificant compared with total energy, i.e. the temperature of the drop is the same before and after the event. In the nuclear case the temperature of B is substantially lower than that of C, and it is the *former* temperature that characterizes the energy distribution of particles Q.

* κ is a dimensionless constant, ζ is a constant with dimensions of energy.

To obtain a quantitative estimate for $P(E)$ we need, of course, to know how $\rho(E)$ varies with E, or what amounts to the same thing, a functional relation between T and E. It is convenient to use this latter formalism, since for nuclei which are composed of many nucleons the statistical treatment is likely to be reasonably valid, and the single parameter T provides a direct indication of the shape of the energy spectrum of the emitted particles provided the approximations mentioned, $\sigma(W_Q)$ constant and high-level density in B, are justified. We shall return later (§ 21) to the question of the dependence of T on the excitation energy E. For the present purpose it will suffice to state that T is approximately 1·0 MeV. for $Z \sim 60^*$ and an excitation energy, E, of 10 MeV., and that T increases with E approximately as $E^{\frac{1}{2}}$ and decreases slowly with A (since the excitation energy is shared amongst a larger number of nucleons) at a rate of between $A^{-\frac{3}{2}}$ and $A^{-\frac{1}{2}}$, more precise values requiring the assumption of a specific nuclear model.

For light nuclei ($Z < 20$) the spacing between levels is probably at least a few kilovolts for the highest bound states, which is not so small that the individual levels cannot be resolved experimentally. Nor is the statistical treatment likely to represent a good approximation in such cases. It is more appropriate then, in discussing reactions involving these nuclei, to exhibit the variation of intensity of the individual 'lines' of the energy spectrum of the particles Q, rather than the number of particles in a given energy interval. Fig. 2a (solid curve) shows the general trend to be expected for uncharged particles, although there may be large fluctuations from level to level in any particular case, and there will probably also be a general tendency for $\sigma(W_Q)$ to decrease for large values of W_Q which will partly offset the factor W_Q in (2·2).

For heavier nuclei the distribution in energy for particles Q may be represented as a continuous curve by (2·5). Fig. 2b, c (solid curves) illustrate the general trend in this case for uncharged particles. It must be borne in mind that, particularly at the high-energy (W_Q) end of the spectrum, the distribution will consist of discrete lines. The decrease in $N(W_Q)$ for large W_Q is due to the increased *spacing* between levels and not to the decrease in the probability of excitation of an individual level.

* κ can be put numerically equal to unity and temperature and energy measured in the same units.

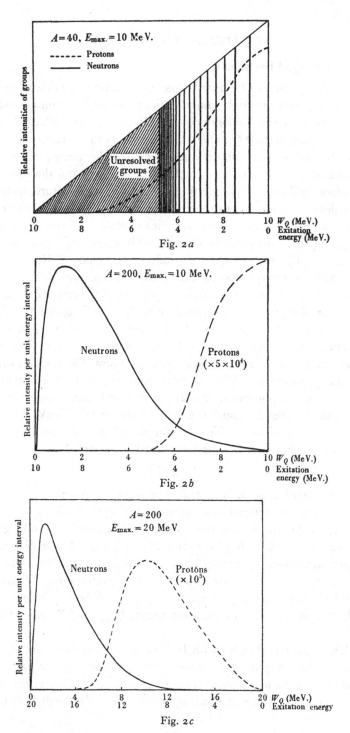

Fig. 2. Average spectrum of particles from compound nucleus.

2·2. Charged particles [4]

If the particles Q emitted from the compound nucleus carry a charge, then the above considerations require substantial modification, since the Coulomb repulsion between the nucleus B and the escaping particle Q imparts a large energy to the latter. Except in the case of the lightest nuclei ($Z < 10$) the energy normally acquired by a proton or α-particle in escaping from the nuclear surfaces will be much larger than the nuclear temperature, so that the distribution functions shown in fig. 2 will be strongly distorted, especially for large excitation energies of nucleus B (small energy W_Q). In the expressions (2·2) and (2·5) it is no longer possible to regard $\sigma(W_Q)$ as even approximately independent of W_Q, but to the same approximation as previously we can regard σ as the product of two factors; a Coulomb penetrability factor $G(W)$ and a factor $\eta(W)$ which is independent of W. $G(W)$ does not reach unity when W is equal to the 'height' of Coulomb-potential barrier, since this quantity refers to the maximum effective repulsion potential for particles of zero angular momentum. For particles with angular momenta l different from zero, there is an additional 'centrifugal' repulsion equal to $l(l+1)\hbar^2/2mr^2$ which amounts to an increase in effective barrier height of the order 1 MeV. for an increase in l of unity. The maximum value of l that need be considered, at any energy, is determined by the relation

$$l\lambda \lesssim 2\pi R. \qquad (2\cdot7)$$

This relation must be satisfied if there is to be appreciable interaction between an incoming and outgoing wave corresponding to particle Q, with angular momentum l and de Broglie wave-length λ, and the nucleus B with radius R. As a typical example: for a proton with energy 10 MeV. λ is approximately 10^{-12} cm., and if we adopt the value $R = 1\cdot5 \times 10^{-13} A^{\frac{1}{3}}$ cm., then the value of l given by the equality in (2·7), denoted by $l_{(\text{max.})}$, will be approximately $A^{\frac{1}{3}}$. For an α-particle of the same energy $l_{(\text{max.})}$ will be twice this value.

The variation of $\sigma(W)$ with W is most marked for values of W smaller than the potential barrier. Thus, for example, with protons and the calcium nucleus, ($Z = 20, A = 40$), $\sigma(W)$ reaches an almost constant value (denoted by $\sigma(\infty)$) when W exceeds 8 MeV., is about $0.6\,\sigma(\infty)$ when $W = 5.8$ MeV. which is equal to the barrier height,* but has dropped to about $0.09\,\sigma(\infty)$ when $W = 2.9$ MeV. For protons and tantalum ($Z = 73$, $A = 181$) the corresponding figures are $\sigma(W) \approx \sigma(\infty)$ for $W > 20$ MeV., $\sigma(W) = 0.3\sigma(\infty)$ for $W = 13$ MeV. (the barrier height) and $\sigma(W) = 0.002\,\sigma(\infty)$ for $W = 6.5$ MeV. For α-particles due to the smaller value of λ corresponding to energies, W, of the order of the barrier height, the increase in $\sigma(W)$ when W *exceeds* the barrier is more marked than in the case of protons but is still not more than a factor 10 even for the heaviest nuclei. The main difference then, between the distribution in energy of charged and uncharged particles will be manifest for particle energies less than the barrier height. From the above examples we can also see that the distribution in energy will be rather different for light and heavy nuclei if the maximum excitation energy is the same in the two cases. In fig. $2a$ the dotted curve shows the distribution in energy for emitted protons ($E_{\max.} = 10$ MeV.) leaving a residual nucleus Ca^{40}. This curve represents only an average trend in the variation of intensity of the discrete groups, larger fluctuations from level to level being likely in the case of charged particles than uncharged ones. This is because in evaluating the variation of $G(W)$ with W, an average is taken over all effective values of angular momentum l, whereas for the production of nucleus B in a particular excited level only a restricted range of angular momenta may be involved, and the actual values may vary from level to level. This factor can be of particular importance if the compound nucleus C can only be produced in states within a small range of angular momentum values on account of the low energy of the *incident* particle P.

In fig. $2b$, which is drawn for protons and $A = 200$ with $E_{\max.} = 10$ MeV., the effect of the potential barrier is to produce a maximum in the distribution at an energy W, approximately equal to the barrier height (or at $W_{\max.}$ if the latter is less than the barrier height). Fig. $2c$ shows the same case but with $E_{\max.} = 20$ MeV.

There are important limitations to the applicability of the general arguments on which the type of results illustrated in Fig. 2 are based. Denoting a nuclear reaction in the style (2·1) is more than

* The barrier height in MeV. is equal to $ZA^{-\frac{1}{3}}$. The accuracy of this expression is about as good as the precision with which the quantity can be defined.

a manner of writing. It indicates that the assumption can be made that an intermediate nucleus is formed in a state, or superposition of states, whose properties depend on the energy of C and not explicitly on the manner of its production. This assumption may be justified in two extreme circumstances: the first when the separation of the levels of C is much larger than the level width so that C is formed in a single well-defined state; the second when there is a large overlapping of the levels so that C is formed in a superposition of many states randomly related in phase. In some type of nuclear reaction a compound nucleus, in the above sense of the term, is not formed (e.g. (d, p) reactions and inelastic scattering discussed below), and then the arguments that have been used to estimate the spectra of emitted particles are inapplicable. In the case where a compound nucleus is formed in a single definite state, or the superposition of only one or two states, there will be factors which may seriously modify the general trends shown in fig. 2. Selection rules deriving from the conservation laws of total angular momentum and parity may restrict both the type of particles which are emitted from a particular level of C, and, more generally, the range of angular momenta (and hence the range of kinetic energies) with which they can be emitted. Where the level spacing of C, at the excitation energies concerned, is large, it will be necessary to consider the properties of individual states rather than the general statistical factors. This is normally the case for light nuclei ($Z < 20$).

2·3. Experimental procedure

The production in a nuclear collision of a nucleus B in an excited state is usually detected in one of two ways: (a) by observation of the kinetic energy of the particles after collision, (b) by detection and measurement of the energy of the γ-radiation when the excited nucleus reverts to its ground state. The half-life for the latter process is, in most cases, large compared with the time taken for the emission of the particle Q from the compound nucleus C (10^{-16} sec. or less), but sufficiently short (usually $\ll 10^{-6}$ sec.) to be regarded as instantaneous from the experimental standpoint. If the energies of the excited states are to be deduced from the measurements of the kinetic energy of the particles Q (the kinetic energy of B is inferred from the conservation of momentum), then it is necessary that the incident particles be homogeneous and the target thin. The actual degree of homogeneity and the target thickness are determined by the available precision of measurement of the kinetic energy of Q, by the spacing of the levels of B that are being investigated, by the requirement of sufficient intensity for accurate measurement, and possibly also by the need to distinguish the particles Q against a large background of particles P elastically scattered by the Coulomb field of the bombarded nuclei or some supporting material. In deducing excitation energies from the observed values of the kinetic energy of Q, allowance must be made for the kinetic energy of B, and this requires that observations of W_Q be made at directions which are well defined with respect to the beam of bombarding particles P. In addition to measurement of the kinetic energies of the particle groups it is necessary to identify the particular group corresponding to the production of B in the ground state. This is usually assumed to be the group of maximum energy that is observed, and this assignation can sometimes be checked from an independent knowledge of the masses of the nuclei involved in the process. Current technique makes it possible to identify this group experimentally by observation of the *simultaneous* (i.e. $\ll 10^{-6}$ sec.) emission of γ-radiation with particle groups of all energies *except* the group corresponding to the nucleus B being produced in the ground state.

Although maximum homogeneity in the energy of the incident beam of particles may be necessary for resolution of the emitted particles into particular groups, the precise value chosen for W_P is not usually important. The energies produced by a cyclotron of about 50 in. diameter suffice for most purposes; and for a study of the bound states of residual nuclei, the relative inflexibility of this instrument as regards changes in bombarding energy is not a serious handicap. It is usually necessary to adopt some special technique to obtain adequate homogeneity in the emergent beam. A series of narrow slits placed in the fringing field of the cyclotron magnet has been successfully employed as an energy filter, although the method involves a large reduction in the intensity of the beam of particles (5). Apart from the dictates of availability, the order of magnitude of the bombarding energy employed may be determined by the dependence of the spectrum of emitted particles on the maximum excitation energy in the manner already discussed and, of course, by the strong dependence of the yield of the process on bombarding energy if this is not equal to or greater than the Coulomb barrier of the bombarded nucleus. An additional factor favouring high energies is the decrease in energy inhomogeneity, introduced by the finite target thickness, with increasing bombarding energy.

In assessing the scope of different methods of exciting and investigating bound nuclear states, it should be borne in mind that no nuclei are known (except He^4) which can have such states at excitation energies exceeding 10 MeV., and our present knowledge of nuclear masses indicate the extreme unlikelihood of such states occurring. Furthermore, an excitation energy of this amount is usually sufficient to dissociate a neutron from the nucleus, and if the neutron can be emitted with an energy of the order 1 MeV. or more, the level width will, except for light nuclei, exceed the level spacing and most traces of a discrete level structure will disappear. (This is not so if only a charged particle can be emitted, since then the Coulomb-potential barrier may reduce the level width by a large factor.) Except then, for light nuclei, the study of all bound states, and of most cases where discrete level structure is in evidence, will be limited to excitation energies (of the product nucleus) of about 10 MeV. or less. In the following discussion of particular processes we shall assume that the bombarded nucleus A is a naturally occurring, and therefore normally β-stable, nucleus. Experiments in which β-active nuclei have been bombarded have not, as yet, been reported.

2·4. Excitation by protons

(p, n) *reactions*: $A(Z, N) + p \rightarrow B(Z + 1, N - 1) + n$. Since the mass of the neutron exceeds that of the proton by approx. 1·2 MeV., and since the nucleus B produced in a (p, n) reaction must have a mass greater than $M(A) - 0·5$ MeV. (otherwise A would not be a stable nucleus; it would be unstable against β-decay leading to B), there must exist a threshold energy W_P(min.), corresponding to zero energy of the emitted neutrons, for any (n, p) reaction. The maximum excitation of B will be limited to $E_{\text{max.}} = W_P - W_P$(min.). The threshold energy is usually between 1 and 3 MeV., so that to excite the lower levels of B, say up to 4 MeV., W_P must be about

6 or 7 MeV. For light nuclei this energy is sufficient for the Coulomb barrier to have little effect on the overall yield, but for heavy nuclei ($Z \gtrsim 50$) somewhat larger energies would be required. With $W_P \sim 10$ MeV. and $E_{max}. \sim 8$ MeV., the maximum in the energy distribution of the emitted neutrons will occur at about 1 MeV. For heavier nuclei then, where larger proton energies are required, the excitation of the lower levels of the result nucleus may be very small, but for $Z < 50$ the (p, n) reaction provides a means of investigating the full spectrum of bound levels of β^+-unstable nuclei produced.* Experimentally, the complete investigation of the spectrum would involve, apart from the production of a sufficiently homogeneous proton beam,† the resolution of the neutron spectrum into components with energies of from about 1 to 8 MeV. and spaced perhaps some 1000 eV. or less apart at the lower end of the neutron spectrum. The best techniques so far developed for measuring neutron energies are capable of a maximum resolution of some 30,000 eV. (6,7). Apart, however, from the resolution of individual levels of B, measurement of the *unresolved* neutron spectrum would provide information regarding the variation of the density of levels with excitation energy.

(p, α) *reactions*: $A(Z, N) + p \rightarrow B(Z - 1, N - 2) + \alpha$. With few exceptions—light nuclei of exceptionally high stability—the (p, α) reaction is exothermic. The binding energy of a proton is usually between 6 and 8 MeV., and that of an α-particle decreases with increasing Z from an average value of 8 MeV. ($Z \sim 16$) to approximately zero for $Z \sim 60$ and is negative for still higher Z (i.e. nuclei are unstable against α-emission but with indetectably small probability of decay). For an incident proton of energy W_P, the maximum energy of the emitted α-particle will range from $W_Q = W_P$ ($Z \sim 20$) to $W_Q = W_P + 10$ MeV. ($Z \sim 80$). The potential barrier for α-emission increases from 10 to about 25 MeV. for the same range of Z. For appreciable excitation of even the lower levels of the product

---

* This term includes instability by orbital electron capture.

† Even with a perfectly homogeneous beam of protons and an ideally thin target there would still be a residual broadening of the discrete neutron groups due to the thermal motion of the atoms in the target. The magnitude of this effect is given by the expression $\Gamma(\text{Dopp.}) = \sqrt{\{(4m/M)\, WkT\}}$. For an incident proton energy of 6 MeV., $M/m = 100$ (ratio of the masses of the bombarded nucleus and a neutron), and a target temperature, T, of 400° abs., $\Gamma(\text{Dopp.})$ is 150 eV. (cf. §8·3).

nucleus, W_P will, then, have to be in the neighbourhood of 10 MeV. or more. This energy is, as we have seen, more than sufficient for the (p, n) process to occur, and since then neutrons can be emitted even with low energy, corresponding to the numerous highly excited states of the residual nucleus, whereas only high-energy α-particles corresponding to the few lowest states are likely, the competition between the two processes will result in a very low intensity of the (p, α) reaction. The usefulness of the (p, α) reaction in the study of bound levels is limited to observations on a few light nuclei (e.g. B^{11}, F^{19}, Na^{23}).

(p, p) *reactions*. For proton energies exceeding the potential barrier, re-emission of a proton from the compound nucleus, i.e. elastic or inelastic scattering, may be of comparable probability to the emission of a neutron. For an appreciable range of levels of the bombarded nucleus to be excited, the proton energy corresponding to maximum excitation must be comparable to the barrier height. Thus proton energies of the same order of magnitude as for (p, n) reactions would be required for light nuclei, and rather higher energies (about 15 MeV.) for the investigation of the level structure of heavy nuclei, although in this latter case there would be strong competition from neutron emission. The arguments used in §2·2 can, however, only be applied with reserve in the case of inelastic scattering of charged particles, since it is possible for excitation of the nucleus to occur by means of the long-range Coulomb inter-action between nucleus and bombarding nucleon. This process does not involve the formation of a definite compound nucleus which disintegrates in a manner independent of its mode of production [8].

For light nuclei the problem of detecting the scattered protons is frequently complicated by the emission of α-particles of comparable or greater range or number (e.g. $Li^7(p, \alpha) He^4$). With heavy nuclei the large Rutherford elastic scattering may also be a handicap. For example, with $Rb(Z = 37, A = 85)$, potential barrier 8·5 MeV., the cross-section* per solid angle 4π at 90° for Rutherford scattering of 10 MeV. protons is approximately $1·3 \times 10^{-25}$ cm.^{2}, which is about the same as the total cross-section for formation of a compound nucleus, although the cross-section for inelastic scattering

---

* The actual elastic scattering will be modified by the breakdown of the inverse square law at distances of the order of the nuclear radius.

without formation of a compound nucleus may be much larger (cf. §2·62). The main advantage of the method of inelastic scattering is that it provides direct information (unambiguous if only a single isotope is present in the scattering material) regarding bound states of stable nuclei. No knowledge of nuclear masses is required. Protons are the most suitable particles for such studies, being generally preferable to α-particles on account of the smaller barrier energies and to neutrons because of the difficulties associated with canalization and energy measurement.

Using bombarding energy of 6·9 MeV. Dicke and Marshall (9) have examined the scattered protons, at 135° to the incident beam, from Al, S, Cr and Mg, the fringing magnetic field of the cyclotron producing the bombarding protons being used to resolve the scattered protons into different energy groups with an accuracy of about 0·03 MeV. Wilkins and his collaborators (10), using protons of the same energy, have investigated the scattering at a large number of angles between 25° and 160°, employing the photographic plates for detection and energy measurement and achieving results of comparable accuracy.*

2·5. Excitation by deuterons

Excitation by deuterons normally leads to a highly excited compound nucleus from which the emission of protons, neutrons and α-particles are all energetically possible. For an average case the excitation of the compound nucleus is about 14 MeV., i.e. the sum of the binding energies of a neutron and a proton less the binding energy of the deuteron (2·2 MeV.). The energy available for neutron and proton emission are both roughly equal to 6 MeV. $+ W_P$, and for α-particle emission about 8 MeV. $+ W_P$ for light nuclei increasing to $W_P + 16$ MeV. for heavy nuclei. Except for light elements the energy available for α-particle emission is more than offset by the effect of the potential barrier. The effect of the barrier on the proton emission is compensated by the possibility of the (d, p) reaction occurring without formation of a compound nucleus, in the sense indicated in §2·2, and hence the proton

* Deflection in the magnetic field of the cyclotron magnet has also been used, in recent experiments, to study the energy spectrum of the scattered protons. (Fullbright and Bush, *Phys. Rev.* **74**, 1323 (1948).)

emission is not regulated by the barrier penetrability. The deuteron is 'polarized' and split in the Coulomb field of the nucleus and the neutron is captured without the proton entering the nucleus ('Oppenheimer-Phillips' process(11)). This process, and neutron emission following the formation of a compound nucleus, are the two most probable processes and have comparable cross-sections for deuterons of energy 5–10 MeV.

(d, n) reactions: $A(Z, N) + d \rightarrow B(Z + 1, N) + n$. On account of the high excitation energies resulting from deuteron reactions comparatively modest bombardment energies suffice for all the bound states of the product nucleus to be excited in a (d, n) reaction. We can regard this reaction as equivalent to the capture of a proton with energy $W_P - 2$ MeV. (binding energy of deuteron). Hence if W_P exceeds 2 MeV. the excitation of all bound states of B is energetically possible. For light nuclei $(Z < 20)$ this energy (~ 2 MeV.) will give sufficient yield for the reaction; for heavy elements energies of about 10 MeV. are required. In such cases the maximum excitation energy of the residual nucleus will be $W_P + 8$ MeV. corresponding to a temperature, for a heavy nucleus, of 1·5–2·0 MeV. The maximum in the spectrum of the emitted neutrons will therefore occur at about 1·5 MeV., corresponding to excitation of the residual nucleus B with energy $W_P + 6$ MeV., i.e. about 16 MeV. Thus many of the levels in which B is produced will be unstable against further emission of a heavy particle so that the (d, n) reaction will be accompanied by $(d, 2n)$, (d, n, α), etc., reactions. The (d, n) reaction results in the same final nucleus as proton capture (cf. §9), so that it is of interest to compare the level spectra exhibited in the two cases. Experiments of the latter type indicate a level spacing at energies corresponding to highest bound states of less than 0·1 MeV. for $Z \gtrsim 15$. A complete comparison of spectra in the two cases would therefore require a high degree of resolution in the analysis of the neutron spectra in (d, n) reactions, but as in the case of (p, n) reactions some useful information can be derived without complete resolution.

Many investigations of neutron spectra in (d, n) reactions have been made, especially for light elements. The neutron energies are usually measured by studing the distribution in range of recoil protons or helium nuclei produced in the gas of a cloud chamber or the emulsion of a photographic plate. The latter method is simpler, but the former is capable of higher resolution, particularly at low energies. Typical results are shown in fig. 3.

Both experiments were made with deuterons having energies of only $1 \cdot 0$ MeV., so that maximum excitation of the product nucleus was $11 \cdot 7$ MeV. in the case of Ne^{20} and $5 \cdot 9$ MeV. for O^{15}.

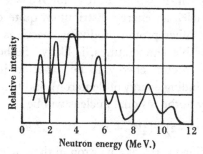

Fig. 3a. Neutron spectra from $F^{19}(d, n)$ Ne^{20}. (Helium recoil nuclei. Spectrum corrected for variation of (n, He) collision cross-section with energy. Bonner. *Proc. Roy. Soc.* A., **174**, 339, 1940.)

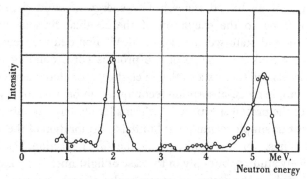

Fig. 3b. Neutron spectrum from $N^{14}(d, n)$. (Proton recoils in photographic emulsion. Gibson & Livesey, *Proc. Phys. Soc.* **60**, 523, 1948.)

(d, p) reactions: $A(Z, N) + d \rightarrow B(Z, N+1) + p$ (12, 13). Just as for the (d, n) reaction, a deuteron energy of 2 MeV. is sufficient, energetically, for all bound states of the residual nucleus to be excited. Transmutations produced by deuterons of this energy correspond to capture of a neutron with zero kinetic energy. A special feature of $(d-p)$ reactions is, as mentioned previously, the possibility of the process occurring without formation of a compound nucleus. This affects both the total cross-section and the energy distribution of the protons produced. For light nuclei

$(Z < 20)$ where the potential barrier is of the same order of magnitude as the binding energy of the deuteron, the 'Oppenheimer-Phillips' process does not play an important role, so that for deuterons of energy 2 MeV. or more, the proton spectrum will be similar to that of the neutrons in the (d, n) reaction.* (The temperature of the residual nucleus will be about 2 MeV. for $W_P = 2$ MeV.) In the case of heavy nuclei, an energy distribution quite different from that observed in other proton-producing reactions with residual nuclei of comparable temperatures, is likely. Bethe [12] estimates that the average proton energy to be expected in this case is $\frac{1}{2}W_P + 2$ MeV. (binding energy of deuteron), and hence the average excitation energy of the residual nucleus will be

$$(W_P + B(n) - 2) - (\tfrac{1}{2}W_P + 2) = \tfrac{1}{2}W_P + B(n) - 4 \text{ MeV.}$$

$(B(n)$ is the binding energy of a neutron in the nucleus B.) This is $4 - \frac{1}{2}W_P$ MeV. less than the energy of the highest bound states of B. Deuterons of energy from 5 to 10 MeV. should be suitable for excitation of all bound states.† The energy of emitted protons corresponding to the excitation of the residual nucleus in the highest bound state will be $W_P - 2$ MeV. For deuteron energies *exceeding* 8 MeV. the range of the protons corresponding to *all* bound levels will then exceed that of elastically scattered deuterons, so that simple range-absorption techniques can be used for investigation of the proton spectrum. With smaller deuteron bombarding energies this simple technique is limited to exploration of the lower bound states. In this limited way the (d,p) reaction has been extensively studied, but only in the case of light nuclei [14]. Experimental study of the proton spectrum is also often hampered by the presence of an intense background of fast neutrons (from the competing (d, n) reaction). When a cloud chamber or photographic

---

* However, in a recent discussion of (d, p) and (d, n) reactions, Peaslee concludes that for the (d, p) reaction the formation of a compound nucleus is less important than the O-P process ('stripping') for *all* deuteron energies, and that stripping is the predominant process even for (d, n) reactions when the deuteron energy exceeds about 10 M.e.v. (Peaslee, *Phys. Rev.* **74**, 1001, (1948).)

† For most nuclei instability against α-particle emission occurs at lower excitation energies than instability against neutron emission. Due to the potential barrier the emission of an α-particle will be an improbable process, so that the width of levels with insufficient energy to dissociate a neutron will be sufficiently small for the discrete nature of the level spectrum to be unaffected.

emulsion is used to record proton tracks, this disturbance is not so serious, since tracks due to recoil nuclei from fast neutron collisions can be distinguished from genuine proton tracks by

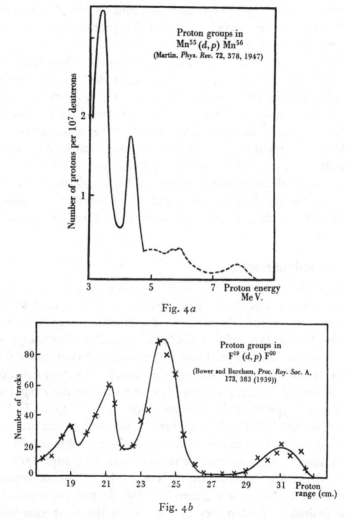

Proton groups in
$Mn^{55} (d,p) Mn^{56}$
(Martin, *Phys. Rev.* 72, 378, 1947)

Number of protons per 10^7 deuterons

Proton energy
MeV.

Fig. 4a

Proton groups in
$F^{19} (d,p) F^{20}$

(Bower and Burcham, *Proc. Roy. Soc.* A, 173, 383 (1939))

Number of tracks

Proton range (cm.)

Fig. 4b

their location and orientation(15). Using range measurements and proportional counters as detectors Martin(16) has employed a pair of collinear counters working in 'coincidence' to reduce the effect of the neutron background. His results for the bombardment of $Mn^{55}(Z = 25, N = 30)$ with deuterons of $3.68\,MeV.$

energy are shown in fig. 4a. The factors limiting the resolution (0·3 MeV.) in this investigation were the energy inhomogeneity of the deuteron beam and the target thickness. Fig. 4b shows the results obtained by Bower and Burcham [17] for the reaction $F^{19}(d,p)F^{20}$, using a cloud chamber for proton detection and energy measurement.

(d, α) reactions: $A(Z, N)d \rightarrow B(Z-1, N-1) + \alpha$. These reactions are only important for very light nuclei, since the Coulomb barrier usually reduces α-particle emission to a very feeble competitor with proton or neutron emission. In cases where it is a reasonably probable reaction, the maximum α-particle energy is between 5 and 10 MeV. plus the deuteron bombarding energy. α-particle groups indicative of excited states in the product nucleus have been observed in the deuteron bombardment of Li, Be, B and N, but in each case the situation is complicated by the fact that the excited states of the product nucleus is unstable against further α-particle emission.

2·6. α-particle reactions

Higher bombarding energies, W_P, are necessary for α-particle transmutations since the Coulomb barrier is twice as large as for protons (barrier 10 MeV. for $Z = 20$ increasing to 25 MeV. for $Z = 80$).* However, owing to the large internal binding energy of the α-particle, reactions produced by it do not result in exceptionally high excitation energies. Thus using α-particles of energy equal to the barrier height the average excitation energy of the *compound* nucleus is approximately constant, being about 18 MeV. for $Z \sim 20$ and 22 MeV. for $Z \sim 80$. This is sufficient for both protons and neutrons to be produced with comparable probabilities. Most of the investigations of the proton and neutron groups emitted under α-particle bombardment were made before artificially accelerated particles were generally available and consequently were limited to light nuclei ($Z < 20$), since the most energetic α-particles available were those from Th C′ (energy 8·95 MeV.) [18].

(α, p) and (α, n) reactions: $A(Z, N) + \alpha \rightarrow B(Z+1, N+2) + p$; $A(Z, N) + \alpha \rightarrow B(Z+2, N+1) + n$. Most of the elements between

* The effective barrier may be slightly smaller due to finite size of the α-particle itself.

$Z = 5$ and $Z = 22$ (the exceptions are C and O) have been disintegrated using α-particles from natural sources (polonium, Ra C', Th C'). The maximum energy of the emitted protons is usually $W_P \pm 2$ MeV. Two or more groups of protons are found in most cases, but the resolution of most experiments is rather poor (~ 0.3 MeV.) on account of low intensity. The (α, p) reactions provide information about many nuclei that can be investigated by the (d, n) and (p, α) reactions. (α, n) reactions have also been investigated for light nuclei. The maximum energy of the emitted neutrons is usually a few MeV. less than W_P (notable exception $Be^9(\alpha, n)$), so that using natural radioactive sources the maximum excitation of the product nucleus is not large. Neutron groups have been found in the bombardment of Be^9, B^{11} and F^{19}.

(α, α) *reactions.* Inelastic scattering of α-particles is governed by considerations similar to those mentioned for proton scattering. The possibility of an inelastic collision occurring without the α-particle penetrating to the nuclear surface is more important in this case, since the penetrability factor will be smaller (for the same ratio of W_P to barrier height). Experiments with α-particles of energy from 11 to 16 MeV. bombarding indium [19] indicate a cross-section for inelastic collision which is very much larger than would be expected if the formation of a compound nucleus were the responsible mechanism, but several orders of magnitude smaller than theoretical estimates of the cross-section for excitation by long-range Coulomb forces [8].

2·7. Neutron reactions

The measurement of the spectrum of particles produced in neutron reactions has only been used in isolated cases as a method of investigation of bound nuclear levels. The difficulties associated with the production of monoenergetic beams of fast (energy of the order of few MeV.) neutrons of adequate intensity limit the range of application of this method. Slow neutron (energies $\ll 1$ MeV.) reactions, other than capture, are important only for light elements. (n, p) reactions, $A(Z, N) + n \to B(Z - 1, N + 1) + p$, using slow neutrons are energetically possible in some cases, but if A is a stable nucleus B will be β^--radioactive, with a mass exceeding A by at least 0·5 MeV. (electron rest mass). Hence the maximum proton energy will be about 0·8 MeV. (neutron – hydrogen atom mass difference) less the energy of the β-transition (the small kinetic energy of the slow neutron is negligible), so that there is little, if any, possibility that B will be produced in other than the ground state. In fact, apart from the possible exception of $N^{14}(n,p) C^{14}$, this reaction has never been observed with slow neutrons.

(n, α) reaction: $A(Z, N) + n \rightarrow B(Z-2, N-1) + \alpha$. This reaction has been observed for most light nuclei from $Z = 3$ to $Z = 30$, but only in a few cases has the energy spectrum of the α-particles been measured. The maximum energy of the α-particles is usually a few MeV. less than the neutron energy, W_P, so that slow neutrons cannot, in general, be used. $Li^6 (n, \alpha) He^3$ and $B^{10} (n, \alpha) Li^7$ are the two notable and much-studied exceptions. Two groups of α-particles have been observed in the boron reaction, using slow neutrons, corresponding to a well-established level of Li^7 at 0·48 MeV. (20). Some inconclusive evidence of two α-particle groups in the reaction $N^{14} (n, \alpha) B^{11}$ has been obtained using fast neutrons (21).

(n, n) reactions. Inelastic scattering of neutrons of energies exceeding 2 or 3 MeV. is a very probable process for all nuclei, but the technical difficulties involved in the detection and measurement of the scattered neutron groups have, so far, prevented the method being utilized.

2·8. Excitation by other particles

Nuclear reactions produced by bombardment with particles heavier than α-particles are not likely to be a very fruitful method of investigating bound levels on account of the high bombarding energy that would be required for adequate yields. The large excitation energies that would occur in such processes would also result in the frequent occurrence of reactions in which more than two groups of nucleons are produced, and the detailed analysis of such reactions is much more difficult than that of the simpler types discussed above. The use of the isotopes of hydrogen or helium of mass 3 may provide a useful extension of technique. The total binding energy of the compound nucleus would be about 17 MeV. $+ W_P$, which is sufficient for the emission of either a neutron or proton and excitation of all the bound states of the residual nucleus. In addition, the nucleus H^3 provides the means of producing very energetic neutrons by the reaction H^3 (d, n) He^4.

3. γ-RADIATION FROM PARTICLE CAPTURE PROCESS AND INELASTIC SCATTERING

In the foregoing account chief attention has been paid to the possibility of production and detection of heavy particles emitted in nuclear reactions. This has been done because the measurement of the γ-radiation emitted by the residual nucleus is usually a less accurate procedure and provides less direct information about the position of excited levels than measurement of the particle spectrum, although γ-energy measurements often provide confirmatory information. In some circumstances particle groups are not emitted, or their examination is impractical, and then examination of the γ-ray spectrum provides an alternative means of investigation. The most important of such cases is the case of capture of the bombarding particle by the bombarded nucleus with subsequent γ-radiation from the compound nucleus

$$A(Z, N) + P(z, x) \to C(Z + z, N + x) + \gamma_1 + \gamma_2 \dots .$$

Re-emission of the incident particle P from the compound nucleus C is always a possible alternative to γ-emission (principle of microscopic reversibility), and in most cases particle emission is very much more probable than γ-radiation. With monoenergetic particles, P, the most intense γ-radiation will occur when the re-emission of particle P is made improbable by the particular circumstances.* For slow neutrons ($W_P \ll 1$ MeV.) this will be so because the re-emission of the neutron in this case is a very improbable process (cf. §8); for charged particles the reduction in probability can arise from the effect of the Coulomb barrier. The emission from C of a heavy particle other than P may be impossible energetically or may be rendered improbable by the operation of

* In practice the intensity of γ-radiation is independent of the probability of emission of P, or any other heavy particle, up to the point where the width of the state in which the compound nucleus is formed does not exceed the range of energies in the incident beam of particles P and provided that the probability of emission of P is not less than that of γ-radiation or any other competing process (cf. §9).

some selection rule or by the effects of the Coulomb barrier. The probability of emission of γ-radiation corresponds to a width of only a few volts or less, and the observed γ-intensity will be greater when the width associated with competitive particle emission from C is small, so that strong γ-radiation arising from capture processes will usually occur from relatively narrow levels of the compound nucleus C. For light elements where the level spacing is not too high, and values of W_P not exceeding 1 or 2 MeV., the levels of C ('resonance levels') can be resolved without much difficulty, the energy homogeneity of the bombarding particles required being 10^3–10^4 eV. With slow neutrons where much greater homogeneity (in absolute magnitude) is possible, discrete levels have been resolved in even the heaviest nuclei. These levels, which are virtual levels of C since they are unstable with respect to emission of P at least, will be discussed in detail in Chapter III. Their significance in connexion with the study of *bound* states arises from the possibility of de-excitation of C by the emission of several γ-quanta in cascade, so that the quantum energies of the γ-radiation will correspond to differences in energy between the excited resonance level of C, the ground state of C and the bound states of C. Apart from the measurement of the γ-ray energies, the most important problem is the determination of the sequence of emission of the several quanta. This information is essential if the information regarding transition energies is to be interpreted in terms of energy levels.

The most extensively studied capture processes are the resonant capture of slow neutrons and capture of protons by light elements. In most investigations emphasis has been placed on the detailed examination of the resonance levels of C rather than on the spectrum of the γ-radiation, and even where attempts have been made to measure the spectrum the results have not been very precise or conclusive (22, 23). For example, the resonant capture of protons with energy 0·44 MeV. by Li^7 results in a virtual state of Be^8 with excitation energy approximately 17 MeV. (In this case the emission of energetic α-particles is forbidden if the assumption is made that the excited state of Be^8 has either odd angular momentum or odd parity.) Despite the fact that γ-radiation emitted for this process is amongst the most intense emitted in proton capture reactions, the exact spectrum of the radiation, apart from identification of a strong line at 17 MeV., has not yet been settled conclusively (24).*

The γ-ray spectra† consequent on the capture of slow neutrons have so far received only scant attention. It is to be expected that with the greatly improved techniques of irradiation by slow neutrons more detailed studies will be made of these γ-spectra which should provide valuable information about the bound states with energies up to several MeV. of a large range of nuclei. In a few instances, when the study of particle groups in a nuclear reaction (as opposed to particle capture) is particularly difficult, study of the γ-radiation has provided the most readily accessible information. Thus in the scattering of protons of energy about 1 MeV. by Li⁷ a small fraction of the protons are scattered inelastically, resulting in an excited state of Li⁷ at about 0·48 MeV. Detection of the inelastically scattered group of protons in the presence of a large intensity of protons of greater energy would be a somewhat elaborate procedure, but the detection of the γ-radiation resulting from the de-excitation of the Li⁷ nucleus from the 0·48 MeV. level is comparatively straightforward (25).

It sometimes happens, both in capture and disintegration processes, that in the course of the de-excitation of the product nucleus a low-lying bound state occurs which has an exceptionally long lifetime (i.e. $\gg 10^{-10}$ sec.), so-called 'isomeric' states (cf. § 17·2). In such cases the γ-radiation can be detected and measured after the bombardment has ceased, a circumstance of considerable technical advantage. The radiation can usually be distinguished from the radiation which may follow the β-activity of nuclei produced by alternative energetically possible processes by its characteristic decay period, its energy, or by establishing the value of the charge on the nucleus responsible for its emission. There is particularly little ambiguity in assignment when the incident particles have too small an energy to cause a nuclear disintegration, but the energy is sufficiently large to make the process of radioactive capture improbable, e.g. the excitation of isomeric state of In¹¹⁵ by the inelastic scattering of neutrons (26), protons (27) and

* A component of about 14·5 MeV. energy has now been definitely established. (Walker and McDaniel, *Phys. Rev.* **74**, 315 (1948).)
† This γ-radiation must be distinguished from any associated with the β-active nuclei that are frequently produced by slow neutron capture.

α-particles (19). It is likely that the isomeric state in these cases is formed by partial de-excitation of a higher combining excited state of the nucleus left after inelastic scattering. In general, then, radiation from an isomeric state can be used as an indicator to measure the extent to which higher levels have been excited in a particular process.

4. ENERGY LEVELS EXCITED IN α-PARTICLE RADIOACTIVITY

4·1. α-particle 'fine structure' $A(Z, N) \rightarrow B^*(Z-2, N-2) + \alpha$; $B^*(Z-2, N-2) \rightarrow B(Z-2, N-2) + \gamma$

Strictly speaking all states, including the 'ground' state of natural α-radioactive nuclei, should be classified as virtual nuclear levels since they are unstable with respect to emission of a heavy particle. But since the half-lives of natural α-emitting nuclei are very much greater (by a factor of 10^8 or more) than the half-life normally associated with particle emission from an excited nucleus, it is convenient for most purposes to regard the low-lying excited states of such nuclei as effectively 'bound' states, and to regard the more highly excited states of such nuclei as virtual when the half-life for emission of a heavy particle is of the order 10^{-15} sec. or less. However, for a comparison of the emission of α-particle groups from the ground state of a particular radioactive nucleus with the particle groups emitted in nuclear reactions, we may regard the α-particle emitting nucleus as analogous to the compound nuclei formed in nuclear reactions. Many of the considerations applicable in the latter case will apply to the former if we bear in mind the great difference in the time scale for the two processes, and the cause of this difference: the small penetrability of the potential barrier of these heavy nuclei. Thus the α-emitting nucleus, A, on account of its long half-life, exists initially in a well-defined state of width quite negligible compared with the level spacing (half-life of 1 sec. corresponds to a width of approximately 10^{-15} eV.), and this state will have completely defined total angular momentum and parity. If the emission of an α-particle leaves the residual nucleus, B, in an excited state, this, too, will normally be a discrete level of B with definite total angular momentum, and hence the wave function describing the emitted α-particle will correspond to a limited range of angular momenta determined by the usual quantum-mechanical rules for addition of angular momenta. We must contrast this situation with that prevailing in the disintegration of a highly excited compound nucleus which we consider as formed in a superposition of many broad levels covering a range of angular momentum values.

Hence the relative intensities of α-particle groups of different energy, although depending primarily on the variation of barrier penetrability with energy, are also influenced by the differences in angular momenta between emitting and residual nuclei, since the values of these quantities determine the angular momentum of the emitted α-particles and this in turn modifies the barrier penetrability (28).

The probability of emission of a particular α-particle group can be considered as the product of two factors: the penetrability function which is dependent on energy and total angular momentum difference between initial and final nuclei, and a factor representing the 'internal probability', the magnitude of which has been the subject of some guesses but little more.

Neglecting the influence of angular momentum, the dependence on energy is represented approximately by the expression for the partial width of the level for emission of α-particles of energy E,

$$\Gamma(E) = C\,\mathrm{e}^{-\delta E^{-\frac{1}{2}}}, \tag{4.1}$$

where δ is a constant for a particular nucleus and depends only on Z, but C ($\sim \mathrm{e}^{-\beta\sqrt{R}}$) involves a knowledge of the radius R of the residual nucleus and is quite sensitive to the value of this quantity (a change of 5 % in R alters C by a factor of 10). C also includes the unknown internal probability factor. The magnitude of δ (for $Z \sim 90$) is such that for $E = 6$ MeV., Γ decreases by a factor of 10 for a decrease in E of 0·2 MeV. Hence α-particle groups corresponding to an excitation energy greater than, say, 1 MeV., will be of extremely low intensity. Several studies of these α-particle groups have been made, using semicircular focusing magnetic spectrographs of high resolution, by Rutherford, Rosenblum and their collaborators (29), and more recently by Chang (30). Chang investigated the α-particles from polonium and radium using semicircular focusing with a photographic plate registering individual α-particle tracks as detector. The use of this type of detector enabled spurious tracks to be identified by virtue of their orientation and location in the photographic emulsion, thus reducing the general background arising from scattered α-particles, contaminations, etc., and

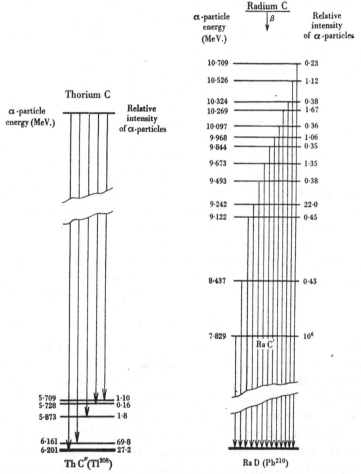

(a) 'Fine-structure' of Th C α-particles.[29] (b) 'Long-range' α-particles from Ra C'

Fig. 5

enabling α-particle groups of very low intensity to be detected. The resolution in energy was about 0·01 MeV. In the case of polonium twelve groups of α-particles were found in addition to the main group of energy 5.305 MeV. (previous investigators had found only this single group), corresponding to emission of an α-particle and the production of Pb^{206} in the ground state. The lowest energy α-particle group observed corresponded to an excited state of Pb^{206} at 1·62 MeV., and had an intensity of 10^{-5} of that of the main group. Earlier investigators of α-particle fine structure had not observed α-particle groups corresponding to excitation energies greater than about 0·5 MeV., and did not report groups with smaller energies than the main groups whose intensities were much less than one-hundredth of the total. Even when all precautions are taken it is quite difficult to establish conclusively the presence of weak α-particle groups whose energy is somewhat *less* than that of a much more intense group.

The α-particle fine-structure for the disintegration ThC→ThC″ is shown in Fig. 5 a.

The energies of the α-particle groups have in general been correlated with the γ-radiation emitted by the residual nucleus. Measurement of γ-transition energies to an accuracy of about 0·001 MeV. is possible for natural radioactive nuclei, since strong internal conversion electrons from sources of radiation of high specific activity can be studied (cf. Chapter IV). Detailed correlation of the γ-ray measurements with the very weak α-particle groups is difficult on account of the feeble intensity of the former.

In the case of Pb^{206} discussed above, the intensities of the short-range α-particle groups are much larger than would be expected from the measurements of the intensity of the γ-radiation from polonium as well as being larger than would be expected theoretically. In view of agreement between α-particle and γ-ray spectra in the case of earlier measurements the apparent discrepancy in the case of polonium may be due to some unsuspected experimental cause (31).*

4·2. Long-range α-particles: $A*(Z, N) \rightarrow B(Z-2, N-2) + α$

These α-particle groups are emitted, in competition with γ-radiation, from excited states of an α-radioactive nucleus, in which the nucleus has been produced as a result of the process

* A repetition of the measurement of the α-particle fine structure of Po by Wadey, using a similar technique to that of Chang, has failed to confirm Chang's results. It appears that the presence of the α-particle groups of low energy is related to the material on which the Polonium is deposited. The presence of low energy particles is due, possibly, to diffusion of Polonium into the supporting material. (Wadey, *Phys. Rev.* **74**, 1846 (1948).)

immediately preceding the α-emission in a radioactive sequence. Since γ-radiation is normally a much more probable process than radioactive α-decay, long-range α-groups are only observed from the shortest lived α-emitting nuclei (Ra C′, Th C′) and then only from excited states of energies 1–3 MeV. above the ground state, resulting in α-particle groups several MeV. more energetic than the main group (fig. 5 b). The low intensity of the *high*-energy groups in this case is due to competition of γ-radiation with α-emission, the former being more probable in most instances by a factor of about 10^6. Since in this process, in contrast to that resulting in α-particle fine structure, all the α-particle groups correspond to the same residual nucleus (Ra D, Th D), the use of the theoretical expression for the variation of emission probability may be more reliable; using this expression, and the measured half-life of the normal α-emission, the absolute probability of long-range α-particle emission is obtained. Measurement of the intensities of the long-range α-particle groups and of the competing γ-rays then provides an estimate of the absolute probability of γ-emission from the excited states. The correlation of the long-range α-particle groups with γ-transitions is relatively easy, since the γ-radiation (associated with the β-decay of Ra C, Th C) is very much more intense than the weak α-particle groups (32).

5. EXCITATION BY β-DECAY

The long-range α-particle groups discussed in the previous paragraph represent transitions from excited states of nuclei produced as a result of β-decay of the parent nucleus (Ra C, Th C). Normally, the excited states resulting from such processes are strictly bound states, i.e. heavy-particle emission is energetically impossible, so that the study of these states is limited to investigations of the β-spectrum or the γ-transitions. The continuous nature of the β-spectrum makes the deduction of information regarding excited levels of the product nucleus less straightforward than in the case of the emission of heavy-particle groups from a compound nucleus. Nevertheless, the study of β-spectra, and the associated γ-spectra, has been extensively pursued to obtain information about excited states, and the method has some very important advantages in its favour: (i) The β-spectrum and the

γ-radiation can be studied without the encumbrances and restrictions associated with the study of processes occurring during nuclear bombardment. (ii) The particular nucleus responsible for the radiations can be identified by the characteristic β-decay period, so that an unambiguous assignment to a definite nucleus can be made even when the original element bombarded contained several unseparated isotopes. (iii) Chemical separation of the β-active material can frequently be made, which not only assists in the identification of the responsible nucleus, but allows sources of radiation of high specific activity to be prepared and to be studied suitably isolated from any material likely to absorb or distort the spectrum of the radiations. (iv) Once produced, the intensity of radiation from a β-decaying material is a well-defined function of time not liable to fluctuations other than the inevitable statistical ones. The main limitations of the method are: (i) Only the lower bound states can usually be investigated, since the *total* energy difference between two neighbouring isobars is normally (light elements excepted) less than 5 MeV. and the maximum excitation energy in a β-decay process usually much less. (ii) For artificially produced radioactive nuclei there are practical limitations imposed by the half-life for β-decay. Too long a life (> 1 year) necessitates long irradiations for appreciable activities; too short a life (< 1 min.) necessitates a study of the radiations intermittently with activation and consequently a sacrifice in some of the above-mentioned advantages. (iii) Most of the readily obtained information concerns the levels of stable nuclei (i.e. non-β-radioactive nuclei).

5·1. Analysis of β-spectra

β-Spectra are divided into two classes, 'simple' and 'complex', according to whether the β-transition always occurs from a single state (usually the ground state) of the initial nucleus to the same state of the final nucleus, or β-transitions occur to more than one state of the final nucleus (fig. 6). In the latter case there will be definite ratios for the numbers of β-transitions to different states, representing the relative probabilities of these transitions, and the observed β-decay half-life will be determined by the sum of these probabilities. All the components of the complex spectra decay with this same half-life.

A third type of β-spectrum arises from β-transitions from two different states of the initial β-active nucleus to either the same or more frequently different states of the final nucleus. Since β-decay is normally a slow process compared with γ-radiation, β-decay from an excited state will only occur with measurable intensity if the competing γ-transition is particularly improbable, e.g. in the case of isomeric states (fig. 6). This type of compound β-spectrum arises really from the superposition of two independent β-spectra, each of the two β-transitions having its own characteristic half-life. The combined spectrum will have a definite form only if the system is in radioactive equilibrium.

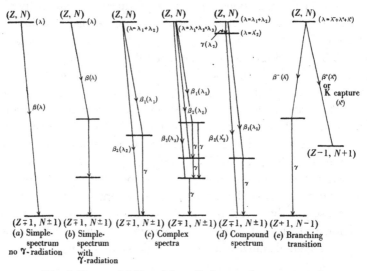

Fig. 6. Types of β-Transition. (λ denotes decay constant.)

Experimental investigation of a β-transition usually proceeds on the following lines: (i) Identification of the β-spectrum as simple or complex from the shape of the spectrum and from (β-γ) coincidence measurements (see below). (iii) Determination of the transition energy for a simple spectrum, and the individual transition energies and relative intensities, and hence partial decay constants, for a complex spectrum. (iii) Measurement of the energies and intensities of the components of the γ-spectrum. (iv) Identification of the sequence of β- and γ-transitions. The analysis of the β-spectrum is based on Fermi's theory of β-disintegration (33), which predicts, for the probability $P(W)$ of the emission of a β-particle with energy W (including rest energy $m_0 c^2$), the value

$$P(W)\,dW = G\,|\,M\,|^2 F(Z, W, R)\,W(W^2 - 1)^{\frac{1}{2}}\,(W_0 - W)^2\,dW, \quad (5\cdot1)$$

where W_0 is the maximum energy of the β-spectrum, $F(Z, W, R)$ is a slowly varying function of W depending on the nuclear charge Z and radius R, G is a constant, determined empirically, representing the strength of the interaction between nucleons and electron-neutrino field, and $|\,M\,|$ is a matrix element involving only nuclear wave functions (i.e. independent of W). Expression $(5\cdot1)$ is based on an approximation in which only the first term in the expansion of the matrix elements of the perturbation term, introduced into the Hamiltonian for a nuclear state to represent the β-transition, is retained. This approximation is justified by noticing that each successive term in the expansion decreases by a factor $2\pi R/\lambda$ (λ is the de Broglie wave-length associated with an electron of energy W_0).* An additional approximation consists in neglecting that part of the perturbation matrix elements which contains the factor v/c (or higher powers), since the velocity v, of the nucleons inside the nucleus, is judged to be only of the order $0\cdot1c$. The operator representing the perturbation has to be chosen so that only those transitions can occur for which the total angular momentum and the parity of the complete system (β-particle plus neutrino and nucleus) are conserved. Selection rules resulting from these conservation laws may cause the first term in the expansion of the perturbation matrix element to vanish. This will be the case unless definite relations are satisfied between the symmetry (spin and parity) properties of the initial and final nuclear states. These symmetry requirements will be different for different types of perturbation operator postulated to represent β-decay; the two most frequently used are the original form due to Fermi (polar-vector operator) and the later development of Gamow and Teller (tensor operator). β-transitions for which the symmetry properties of initial and final states correspond to a non-vanishing first term in the expansion of the matrix element are termed 'allowed'; when

* For heavy nuclei the factor is different on account of the distortion of the electron wave function by the nuclear charge (33).

the second is the first non-vanishing term, they are referred to as 'first forbidden' and so on. The table shows the symmetry requirements (i.e. the selection rules) for the transitions corresponding to non-vanishing terms in the expansion up to the third.

	Fermi		Gamow-Teller	
	ΔJ	Parity	ΔJ	Parity
Allowed	0	=	0, ± 1 (except 0→0)	=
First forbidden	0, ± 1 (except 0→0)	$\neq$	0, ± 1, ± 2	$\neq$
Second forbidden	± 1, ± 2	=	0→0, ± 2, ± 3	=

Selection rules in β-decay. ΔJ is the difference in total angular momentum of initial and final nuclei. $=(\neq)$ indicates that initial and final nuclei have the same (different) parity.

First forbidden transitions should be about $10^{-2} (\approx (2\pi R/\lambda)^2)$ as probable as allowed transitions and second forbidden transitions about 10^{-4} as probable. Measurement of the decay constant (or partial decay constants in the case of complex spectra) should provide, therefore, useful information about the symmetry properties of the initial and final states. Inferences of this type can only be made reservedly for two reasons. First, there is an arbitrariness in the choice between Fermi and Gamow-Teller selection rules, or possibly other selection rules arising from other postulated interaction operators which require different symmetry relations. Secondly, symmetry properties based on approximations regarding the *nuclear* Hamiltonian may lead to variations in the nuclear matrix element which are comparable with those introduced by the absolute conservation laws for momentum and parity, i.e. the probabilities of individual transitions having the same degree of forbiddenness, according to the above rules, may vary by as much as a factor 100 (allowing, of course, for any difference in the energies) (33), (34).

The decay constant for a particular transition is proportional to $\int_0^{W_0} P(W) \, dW$, and its dependence on W_0 is sufficiently great for the transition to a highly excited state to occur with measurable intensity in competition with the transition to the ground state of the product nucleus, only if the latter transition is more highly forbidden than the former. Hence in the case of complex spectra it is generally to be expected that at least one component will correspond to a forbidden transition. For forbidden transitions the expression (5·1) is not strictly applicable, but for large enough Z ($\frac{1}{2}\alpha Z > 2\pi R/\lambda$, which is true for $Z > 10$) all first forbidden spectra should have the

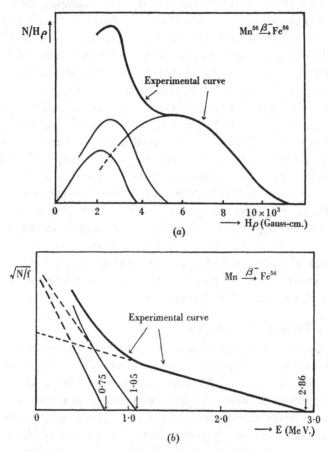

Fig. 7. β-spectrum of Mn^{56}. (a) Spectrum; (b) 'Fermi' plot. Showing analysis into component spectra. (Elliott and Deutsch, *Phys. Rev.* **64**, 322, 1943.)

same shape as allowed spectra. Of the second forbidden spectra, those corresponding to $\Delta J = \pm 2$ give the greatest departure from (5·1). In the analysis of complex spectra it is usual to assume, in the absence of any more reliable procedure, that (5·1) is valid for all the component spectra. A convenient way of representing the observed

β-spectrum, first used by Kurie *et al.*, and now customary, is to plot the quantity $K(W) = [N(W)/(W^2 - 1)^{\frac{1}{2}} WF(Z, W)]^{\frac{1}{2}}$ against W. ($N(W)$ is the number of electrons per unit energy interval.) According to (5·1) the relation between $K(W)$ and W is linear with an intercept on the W-axis equal to W_0. For a complex spectrum, assumed to be a superposition of spectra of type (5·1) each with its characteristic value of $|M|^2$ and W_0, the relation between $K(W)$ and W will consist of a number of linear sections from which the individual values of W_0, and, using the measured half-life, the partial decay constants, can be readily deduced. Fig. 7 shows such an analysis of a typical complex spectrum, one of several studied by Deutsch and his collaborators using a 'short' magnetic lens spectrograph (35). This case is of special interest, since the positron decay of Co^{56} results in the same residual nucleus, Fe^{56}, and so offers an opportunity of comparing the spectrum of levels of the same nucleus excited in two different processes. The values obtained for W_0 for the three components of the Mn^{56} β-spectrum are 0·75 ± 0·1, 1·05 ± 0·03 and 2·86 ± 0·05 MeV., with relative intensities 15:25:60 respectively, indicating energy levels of Fe^{56} at $E + 2·11$, $E + 1·81$ and E MeV., the last corresponding to the partial spectrum of maximum energy (2·86 MeV.) but not necessarily to the production of Fe^{56} in the ground state.

5·2. γ-Transitions in β-decay

Measurement of the γ-spectrum usually shows γ-transitions corresponding to transitions between some pairs of levels whose excitation is indicated from the analysis of the β-spectrum. (γ-Transitions between such levels may also occur in two or more steps, i.e. via additional intermediate levels which do not correspond to the end-state for any β-transition. On account of the close connexion between the selection rules for β- and γ-transitions this situation seldom arises.) Measurements of the energy of the γ-transitions accompanying β-decay is usually somewhat more accurate than measurement of W_0 for the component spectra, so that the results of the former can be used to check and improve the latter (cf. Chapter IV). In the case of Fe^{56} quoted above, γ-transition energies of 0·845 ± 0·015, 1·81 ± 0·04 and 2·13 ± 0·04 MeV., with relative intensities 100:30:20, were measured. γ-Transition energies associated with β-transitions having simple spectra, or those not corresponding to direct γ-transitions between levels excited by component β-transitions in case of complex spectra (e.g. the γ-transition of 0·845 MeV. in the above example), must involve additional states, including possibly the ground state of the residual nucleus. Where there are a number of such uncorrelated γ-transition energies, the problem of inferring the level spectrum is a more involved affair, and cannot, in most cases, be deduced from measurements of β- and γ-energies alone.

5·3. Coincidence measurements

A most important adjunct to the measurement of β- and γ-energies and intensities, one extensively developed in recent years, is the technique of observing 'coincident' emission of a β-particle and γ-quantum, or of several γ-quanta (36, 37). In most cases where a γ-transition follows a β-transition, or a preceding γ-transition, the second process occurs within 10^{-7} sec. or less of the first, so that experimentally the two, or more, processes in succession are registered as simultaneous. The maximum time interval between two transitions for these to be registered as 'coincident' (the resolving time) is normally made as small as technically possible ($\sim 10^{-7}$ sec.) so as to reduce to a minimum the number of random 'coincidences' arising from uncorrelated events. Coincidence measurements can be used to confirm the analysis of a β-spectra by observing the variation in the number of β-γ coincidences as a function of β-particle energy, different parts of the β-spectrum being separated either by absorption or resolution in a magnetic spectrograph. When the β-spectrum is complex the number of β-γ coincidences will vary with β-particle energy and conversely. The occurrence of (β-γ) coincidences for β-particles of maximum energy indicates that no β-transition occurs to the ground state of the product nucleus (fig. 6 b, c).

In the case of simple β-spectra the presence of several γ-energies indicates successive γ-transitions occurring after each β-transition, but there may be several competitive modes of de-excitation of the excited nucleus produced by the β-transition, which means a certain arbitrariness in the possible level spectra which can be deduced from the measured γ-energies. Any postulated scheme of transitions must also agree with the measured *intensities* of the γ-transitions (not usually measured more accurately than about 10 %), and can be tested further by measurement of the number of γ-γ coincidences. This last measurement is often decisive in the choice of one number of otherwise equally plausible level schemes, e.g. in association with the β-decay of Na^{24} two γ-transitions with energies 2·76 and 1·38 MeV. and approximately equal intensities are observed. β-γ coincidences are independent of β-energy. Either of the level schemes shown in fig. 8 a, b are consistent with these results, but measurements of the γ-γ coincidence rate shows scheme (a) to be the correct one (38), a result which also leads to a more reasonable

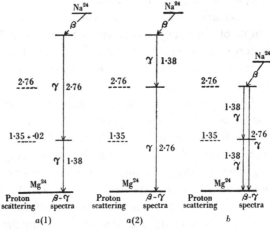

Fig. 8. Level schemes for Mg^{24}. (Energies in MeV.)

result for the mass difference of Na^{24} and Mg^{24}. Interpretation of the measured number of γ-γ coincidences requires a knowledge of the efficiency of the γ-radiation counter used and, in particular, the variation of efficiency with energy (23,36,39). An extension of the γ-γ coincidence technique in which the number of γ-γ coincidences is measured with several counters of different efficiencies has been used by Wiedenbeck (40) to obtain additional evidence in favour of scheme (a) in fig. 8.

In making inferences from the observed number of β-γ and γ-γ coincidences it is usual to assume that the angular distribution of γ-radiation from each excited level is isotropic, i.e. that there is no angular correlation between successive transitions. Theoretical examination of this point by Hamilton [41] and Goertzel [42] indicates that the correlation factor will usually be small for a bare nucleus and will be substantially reduced by interaction with the extra-nuclear electrons if the half-life for emission of the second quantum is comparable with or greater than the period of the precession of the nuclear moment about the resultant atomic moment. (From measurements of optical and radio-frequency spectra, this period is of the order 10^{-9} to 10^{-11} sec.) In two cases (Co^{60}, Sc^{46}) angular correlation of successive quanta has been observed; in several other investigations the effect was either absent or too small to be detected [43].*

It should be noted that if the half-life for a γ-transition following either a β- or a γ-transition is much greater than 10^{-6} sec., β-γ and γ-γ coincidences would not be observed with normal resolving times employed. The use of much longer resolving times would necessitate working with much weaker sources to avoid an overwhelming number of 'random' coincidences.

5·4. Excitation by K-capture

In one type of β-transition, capture of K (or L, M, etc.) orbital electron and emission of a neutrino, since only one particle is emitted its spectrum must consist of discrete lines, each line corresponding to a transition to a particular level of the product nucleus. Unfortunately, the energy spectrum of these particles is not subject to direct examination. Information about the level spectrum of the product nucleus in such cases can be obtained from measurements of the γ-spectrum together with any information regarding the nuclear mass that may be available. Recently there has been observed some weak continuous γ-spectrum ('Bremsstrahlung') associated with the change in 'electron configuration' caused by K-capture (i.e. capture of electron from a bound energy level). The maximum energy of this γ-spectrum corresponds to the neutrino energy of most transitions (only in a small number of the transitions is appreciable energy carried away by this Bremsstrahlung).

* More accurate experiments in which not only the correlation between directions of emission, but also between the planes of polarization of successive γ-transitions, have recently been made. (Brady and Deutsch, *Phys. Rev.* **74**, 1541 (1948), Deutsch and Metzger, *ibid.*, p. 1542.)

K-capture and positron emission are often alternative, competitive processes. The energy available for K-capture is $2m_0c^2$ (1 MeV.) greater than that for β^+-emission, so that in cases where both processes are possible, K-capture will often take place to an excited state of the residual nucleus to which β^+-transitions are energetically impossible. β^+-transitions to excited states very rarely occur.

6. EXCITATION BY γ-RADIATION

There are two types of process by which γ-irradiation can result in nuclei being excited in bound states. In the first type the γ-energy is sufficient to dissociate a heavy particle from the nucleus and leave the product nucleus either in the ground or in an excited state. In the second type, the γ-energy is insufficient for dissociation; the irradiated nucleus can absorb a γ-quantum by making a transition to a bound state, and then revert to the ground state by a single or by a succession of γ-transitions.

6·1. (γ, p) (γ, n) reactions

In order that processes of the first type may give informatoin about the energy levels of the residual nucleus, the γ-energy must be of the same order of magnitude as the excitation energy of the compound nucleus in a nuclear reaction initiated by particle bombardment, and the same considerations of energy homogeneity apply. The only practical method, at present available, of producing homogeneous* γ-radiation of sufficient energy is by employing γ-radiation excited in a nuclear reaction. Capture of protons by Li[7] provides γ-radiation of 17 MeV. energy which is very suitable for many investigations of the type being considered, provided other γ-ray components are not present in the spectrum in disturbing amounts. The most probable process for practically all nuclei excited by 17 MeV. is neutron emission [44] (the average maximum energy for neutron emission is about 10 MeV.). For light nuclei the emission of a proton or α-particle should be of comparable probability. Since the *momentum* of the γ-quantum is extremely small,

---

* The homogeneity of the γ-radiation from the Li[7] (p, γ) process is ultimately governed by Doppler broadening (cf. §8·5). If the incident protons are unresolved the spread in γ-energies will be equal to the width of the resonance level, i.e. about 0·012 MeV., neglecting other pcssible components in the γ-spectrum.

compared with that of a nucleon resulting in the same excitation energy, the energy of the emitted particles in the reactions (γ, n) and (γ, p), etc., should be practically independent of the angle between the direction of emission of the particle and that of the incident γ-radiation. This factor might be utilized to offset partly the small cross-section (10^{-27} to 10^{-29} cm.^{2}) for these processes. Apart from the limited intensity of γ-radiation available and the small cross-sections for (γ, p) reactions, there is another difficulty in the study of particle groups from this process, namely, that of distinguishing the particles emitted from the substance under investigation and those from neighbouring material. Adequate canalization of the γ-radiation would impose too drastic a reduction in intensity. All nuclei except H^1 and He^4 can be disintegrated by 17 MeV. γ-radiation, although C^{12} (carbon is 98·9 % C^{12}), which requires 16 MeV. for dissociation of a proton and 7·4 MeV. for an α-particle, might not result in too large a source of spurious particles.

Many investigations have been made of the β-radioactive nuclei produced in (γ, n) and (γ, p) reactions, but no measurements of the energies of the associated protons and neutrons have been reported [44, 45].

6·2. Excitation of bound states by electrons and X-rays

Direct excitation of bound nuclear levels by electromagnetic radiation occurs only at more or less discrete quantum energies. This follows from the observation* that the level spacing of bound states is greater than the level widths of these states (< 1 eV.) for all nuclei. The ratio of level spacing to level width is probably greater than 10^3 in most cases, so that in the irradiation of matter by inhomogeneous X-rays only a small fraction of the total radiation, that having energy corresponding to transitions from the ground state to an excited state, will interact appreciably with the nuclei.

X-radiation of sufficient energy and intensity for nuclear excitation, which is also homogeneous (i.e. with energy spread small compared with nuclear level spacing), cannot be produced with present-day techniques, so that excitation of bound states has been accomplished using a continuous X-ray spectrum extending up to

* Based on the measurements of slow neutron capture (see §8).

a few MeV. or by irradiation, with beams of high-energy electrons. Interaction between an electron with an energy of a few MeV. and the nucleus can be regarded as essentially the same as that between the nucleus and electromagnetic radiation, since the specifically nuclear interaction between electrons and nuclei is extremely weak (witness the long half-lives for β-decay). Detection of nuclear excitation by observation of the resonant absorption or re-radiation of X-rays (analogous to the processes easily observed in atomic spectra) is extremely difficult on account of the much stronger interaction of the radiation with the extranuclear electrons (photo-electric absorption, Compton scattering and pair-producing processes). For excitation of the nucleus to a level with energy E_r, total width Γ ($\Gamma = \sum_i \Gamma_i$, where Γ_i is the partial width for a particular γ-transition from the excited level) and partial width Γ_r for the γ-transition with energy E_r, the observed cross-section will be given by*

$$\bar{\sigma} = \tfrac{1}{4}\pi\left(\int\{\lambda^2(E)\, g\, \Gamma_r\Gamma/[(E-E_r)^2+\tfrac{1}{4}\Gamma^2]\}\, I(E)\, dE\right)\bigg/\left[\int I(E)\, dE\right],$$

$$(6\cdot1)$$

where $I(E)$ represents the spectrum of the radiation, $\lambda(E)$ is the wave-length corresponding to a γ-energy E, and g is a statistical factor of the order unity depending on the angular momenta of the nuclear states and the multipole nature of the interaction (46). Since $I(E)$ and $\lambda(E)$ are only slowly varying functions of E, compared with the resonance term in (6·1), we can write (6·1) as

$$\bar{\sigma} = \tfrac{1}{2}[\lambda^2(E_r)\, g\, I(E_r)\, \Gamma_r]\bigg/\left[\int I(E)\, dE\right].$$

If we remember that Γ_r is of the order 1 eV. or less and that $I(E)$ may spread over a range of the order 10^5 or 10^6 eV., it is clear that $\bar{\sigma}$ will be smaller than $\lambda^2(E_r)$ by a factor of 10^5 or 10^6.

What might appear to be the most favourable circumstance for direct detection of resonance absorption or scattering, namely, when the γ-radiation employed arises from a transition to the ground state of the same nucleus as that in which absorption is being investigated (e.g. the 1·38 MeV. γ-transition of Mg^{24} being absorbed

* Cf. §7·1.

by magnesium),* does not, in fact, provide a good opportunity for detecting these processes. Although the γ-radiation in this case contains a single sharp line of energy almost identical with E_r, the exact quantum energy of this line will be less than E_r, since a small fraction of the transition energy, δE, is converted into kinetic energy of the nucleus recoiling after the emission of the radiation. Similarly, if the radiation is absorbed by another nucleus, not all the energy is available for nuclear excitation, but the same amount, δE, is expended in nuclear recoil. Thus, in evaluating the expression (6·1) the average cross-section will be approximately equal to the cross-section for $E = E_r - 2\delta E$, not that for $E = E_r$. If $\delta E \gg \Gamma$ the reduction in the cross-section will be very large. (In the example quoted above, Mg²⁴, the recoil energy is 55 eV., which implies an effective reduction in the cross-section, as compared with the value of exact resonance, of about 10⁴ even if Γ is as large as 1 eV.) These considerations must be modified if any other motion of either radiating or absorbing nucleus causes a Doppler broadening of the frequency comparable with δE. Such motion will arise from ordinary thermal agitation and as a result of some nuclear process immediately preceding the γ-transition (e.g. β-decay of Na²⁴ in the example cited). If the Doppler broadening, ΔE, is large compared with δE, then the reduction in the absorption cross-section will be of the order $\Gamma/\Delta E$, which is not so severe a result as that caused by the recoil effect. The anomalous absorption of γ-radiation due to nuclear resonant absorption has not, in fact, been demonstrated experimentally.

The difficulties associated with the direct detection of resonant absorption have been overcome in the methods used by Collins, Wiedenbeck (47) and collaborators, to detect nuclear excitation during irradiation by artificially produced X-rays and electron beams of energy up to 3·2 MeV. In these experiments the occurrence of excitation was detected by the observation of delayed radiation from an isomeric level which is excited in the course of de-excitation of the excited nucleus resulting from absorption of the γ-radiation (fig. 9a). The amount of delayed activity was studied as a function of the maximum energy, $E_{(max.)}$, of the X-ray spectrum produced by arresting the electrons in heavy material or as a function of the

* Cf. §5·3, fig. 8.

energy of the electrons which were used to irradiate a thin sheet of the sample.

The variation of activity with $E_{(\text{max.})}$ in the case of X-ray irradiation is shown in fig. 9b for a typical case: cadmium irradiated with X-rays of energies up to 3·2 MeV. There is evidence of discontinuities in the variation of the activity with energy which are

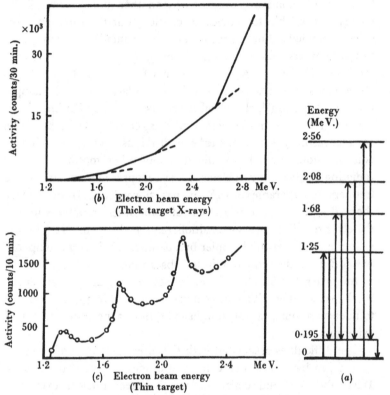

(b)   Electron beam energy
(Thick target X-rays)

(c)   Electron beam energy
(Thin target)

Fig. 9. Excitation of cadmium by X-rays and electrons.
(Wiedenbeck, *Phys. Rev.* **67**, 92, 1945.)

interpreted as occurring when $E_{(\text{max.})}$ is equal to the excitation energy of a particular level. In the case of cadmium these discontinuities occur at energies of 1·25(?), 1·68, 2·08, 2·56 and 2·97 MeV. The activity which was measured was from a 48·7 min. isomeric transition of approximately 0·195 MeV. energy (the conversion electrons were actually detected). This isomeric state had been identified previously by inelastic scattering of neutrons [48].

Similar measurements, with remarkably similar results, have been made with $Ag^{107,109}$, Au^{197} and Rh^{103}. In all cases the first discontinuity occurs at an energy of between 1·18 and 1·26 MeV. and successive ones at regular intervals of about 0·4 MeV.

In the case of electron bombardment, the effective spectrum of radiation which arises from the passage of a fast electron near a nucleus is somewhat different from the X-ray spectrum of the previous experiments. It consists essentially of two parts, one similar to the X-ray spectrum (Bremsstrahlung) which has a maximum intensity at low energies and falls off monotonically with increasing energy, and the other which has a fairly sharp maximum at an energy equal to the kinetic energy of the electron [49]. This second component results in peaks in the excitation curve showing variation of activity with electron energy (fig. 9c).

It must be borne in mind in interpreting the results of these experiments that not all the excited states in the energy range covered will necessarily be indicated by this procedure, since some levels may not result in an appreciable number of transitions to the isomeric level. But in view of the fact that for heavy nuclei (e.g. Au^{197}) the level spacing at excitation energies of 2 to 3 MeV. is certainly much smaller than 0·4 MeV., it is hard to understand the observed constancy in level spacing. Even more surprising is the fact that the same level spacing is observed for nuclei with masses of about 100 and 200.

CHAPTER III

EXCITATION OF 'VIRTUAL' STATES

Virtual levels of a nucleus can be excited directly by bombardment with nucleons or other heavy particles. If a nucleus $C(Z, N)$ has a virtual level with excitation energy E, then this level will be excited as a compound nucleus in the bombardment of the nucleus $A(Z-p, N-x)$ by the particle $P(p, x)$ with energy W_P equal to $E - B(P)$, $B(P)$ being the binding energy of the particle P in the nucleus C. Virtual levels will, in general, have a greater width than bound levels, since particle emission is in general a more probable process than γ-radiation. As the excitation energy is increased beyond the value $B(P)$ the level width will increase rapidly on account of the increased energy available for emission of particle P. In addition, the level spacing *decreases* so that we should expect to find *discrete* virtual levels only for a limited range of energies greater than $B(P)$. For higher energies the overlapping of levels will be sufficient for the effects of individual levels to be smoothed out. If the compound nucleus C is formed with such a high excitation energy, it can be regarded as a superposition of states with a particular set of amplitudes and phase differences depending on the energy and on the nature of the process producing it. If the excitation energy is sufficiently large the number of overlapping levels may be sufficient for the amplitudes and phases to be regarded statistically, and to replace the individual amplitudes by an average amplitude and to treat the phases as random. The properties of such a compound nucleus could then again justifiably be regarded as independent of the particular process by which it is produced and to be only slowly varying functions of energy. But under these circumstances it is only relevant to deal with such average properties as the mean level widths (both total widths and partial widths for a particular process) and the level density, concepts which are of value in interpreting the cross-section for nuclear reactions (cf. §2·2). The following account of experimental methods and results will be limited to the region of excitation energies where the discrete nature of levels is still in evidence, i.e. when the compound nucleus can be described as being in a well-defined state whose properties are not explicitly dependent on its mode of formation.

7. RESONANT PROCESSES

The discrete nature of the energy-level spectrum of the compound nucleus is associated with resonance characteristics in the dependence of the cross-section for nuclear processes on energy. This dependence is represented by the well-known dispersion formula, originally formulated by Breit and Wigner and developed by numerous others (50):

$$\sigma(W)_{(P\alpha l \to Q\beta l')} = \frac{\lambda^2}{4\pi} \left| \sum_r \frac{[\gamma_{P\alpha l}^{Cr}(E) \cdot \gamma_{Q\beta l'}^{Cr}(E)]^{\frac{1}{2}}}{(W - W_r) + \frac{1}{2}i\gamma_{Cr}} e^{i\phi_\beta} \right|^2. \quad (7\cdot1)$$

This expression gives the partial cross-section for the process in which particle P with kinetic energy W, de Broglie wave-length λ (reduced to centre of gravity co-ordinate system), and angular momentum l is captured by a nucleus in a state denoted by 'α' (forming a compound nucleus C), and a particle Q is emitted with angular momentum l' leaving the residual nucleus in state 'β'. $\gamma_{P\alpha l}^{Cr}$, $\gamma_{Q\beta l'}^{Cr}$ refer to the partial widths of the state 'r' of C for emission of P, Q with the appropriate angular momenta, etc., and γ_{Cr} is the total width of the state 'r' of C (i.e. $\gamma_{Cr} = \sum_{Q\beta l'} \gamma_{Q\beta l'}^{Cr}$, where the summation is over all possible processes including emission of P). ϕ_β is a factor representing the phase difference between initial and final states; W_r is the value of W corresponding to exact resonance. All the γ's in the above expression are energy-dependent quantities. When the overlapping of levels is negligible only one level 'r' need be considered, instead of the amplitude summation, and in this case also, the compound nucleus, being in a well-defined state, will have a definite angular momentum J. This permits the summation to be made over the different spin orientations resulting in the 'single-level' formula for the cross-section for the emission of particle Q leaving the residual nucleus in state 'β' (irrespective of angular momenta of P, Q provided that total angular momentum is conserved throughout the process)

$$\sigma(W)_{(P\alpha \to Cr \to Q\beta)} = \frac{\lambda^2}{4\pi} \frac{2J+1}{(2i+1)(2j+1)} \frac{\gamma_{P\alpha}^{Cr}\gamma_{Q\beta}^{Cr}}{(W - W_r)^2 + \frac{1}{4}\gamma_{Cr}^2}. \quad (7\cdot2)$$

J, j, i refer to the total angular momentum of state 'r' of C, and the total (intrinsic) angular momenta of the bombarded nucleus A and particle P respectively.* In this expression the γ's are still energy dependent; $\gamma_{P\alpha}^{Cr}$ refers to the partial width for the emission of P leaving nucleus in state α; similarly $\gamma_{Q\beta}^{Cr}$. These partial level widths are only slowly varying functions of the energies of the incident and emitted particles, so that provided these energies are much larger than the *total* level width, the main dependence of σ on W in (7·2) arises from the term in the denominator. It is often sufficiently accurate to neglect the variation of the γ's for a particular level and replace the actual values by the values at resonance (the γ's will be different for different levels of course).

Measurement of the relative variation of σ with W provides information about the excitation energies and total widths of the virtual levels. Measurement of *absolute* cross-sections provides additional information about partial level widths. Expression (7·2) refers to the total cross-section for the process denoted by $(P\alpha \to Cr \to Q\beta)$, i.e. to the yield of particles Q integrated over all directions of emission. When the overlapping of levels is quite negligible so that (7·2) applies to each resonance independently, then the variation of $\sigma(P\alpha \to Cr \to Q\beta)$ through a resonance will be the same as that of the differential cross-section, $\sigma(\theta)$, for emission of particles in a particular direction, θ (with reference to the incident particles). If there is some overlapping of levels, then due to interference between the terms in (7·1) the angular distribution may vary markedly with energy, and the energy at which the maximum in σ occurs will vary with θ (by an amount of the order γ_{Cr}).†

Pronounced resonance phenomena, indicating level widths much smaller than level spacing, have been observed under conditions of the following types:

(i) The incident and emitted particles are slow. When the wave-lengths, λ, of the particles that can be emitted by the

* It is assumed that the spins i, j are randomly oriented.

† The analyses of the reactions Li⁷ (p, n) Be⁷ and Li⁷ (p, γ) Be⁸ are particularly illustrative of this type of interference. The latter reaction also provides favourable circumstances for checking the detailed predictions of the Breit-Wigner Dispersion Formula. (Breit and Bloch, *Phys. Rev.* **74**, 397 (1948), Devons and Hine, *ibid.*, p. 976.)

compound nucleus are much greater than the nuclear radius, R, the partial widths for the emission of these particles contain a factor R/λ. For slow enough particles ($W \ll 1$ MeV.) this factor reduces the partial width for particle emission to a value smaller than the normal γ-radiation width, so that the latter determines the total width and the predominant process is particle capture (e.g. slow neutron capture, where the level width for emission of *charged* particles is negligible on account of the barrier).

(ii) The level width is reduced due to small penetration through the potential barrier. This factor will be important in charged particle processes in which the emission of a fast neutron is energetically impossible (e.g. proton capture, α-particle disintegration and scattering).

(iii) Reactions involving light nuclei show resonance characteristics, even at high excitation energies (10 MeV. or more), owing to the very large level spacings.

(iv) Emission of a particle with large energy (normally associated with a level width of, say, about 1 MeV.) may not be possible owing to selection rules (e.g. $B^{11}(p, \alpha) Be^8$; $F^{19}(p, \alpha) O^{16}$; $Li^7(p, \gamma) Be^8$).

(v) Bound states, which can only emit γ-radiation, have widths less than level spacings for all nuclei (cf. §6).

8. SLOW-NEUTRON RESONANCE LEVELS[51]

Resonance phenomena in the interaction of slow neutrons with nuclei are amongst the most interesting and thoroughly investigated nuclear processes, and provide direct information about the widths and spacing of virtual levels of most medium and heavy nuclei ($Z > 20$). The reasons for their importance are, first, the absence of Coulomb forces, which means that neutrons, and particularly slow neutrons, can be used to investigate, with the same technique, all nuclei; secondly, since the energy of slow neutrons is very small compared with the excitation energy of the compound nucleus (~ 8 MeV.), even moderate homogeneity of the neutrons enables a small portion of the level spectrum to be studied in great detail; thirdly, the cross-sections for slow-neutron resonance processes are very much larger ($\sim 10^{-21}$ cm.^{2}) than the nuclear size; lastly, the absence of the Coulomb field simplifies the theoretical interpretation of the experimental results. Slow-neutron investigations are limited to the examination of a small energy region of the nuclear level spectrum (up to about 100 eV.), starting at the threshold energy necessary to remove a neutron. In the case of light nuclei, where the level spacing is much larger than this range, the chances are against observing resonance levels by this technique. Fast neutrons (the distinction between slow and fast is largely technical, although the criterion $\lambda \gg R$ is also significant) can be used for similar investigations in light elements.

The application of (7·2) to the case of slow neutrons, gives (remembering $i = \frac{1}{2}$ and $J = j \pm \frac{1}{2}$, since, according to (2·7), only neutrons with zero orbital momentum will be important)

$$\sigma(n, Q\beta) = \frac{\lambda^2}{8\pi}\left(1 \pm \frac{1}{2j+1}\right)\frac{\gamma_n^{Cr}\,\gamma_{Q\beta}^{Cr}}{(W-W_r)^2+\frac{1}{4}\gamma_{Cr}^2}. \qquad (8·1)$$

Since $\lambda \gg R$, the variation of γ_n^{Cr} with neutron energy will be determined predominantly by a factor R/λ, so we can write

$$\gamma_n^{Cr} = (\lambda_r/\lambda)\,\Gamma_n, \qquad (8·2)$$

where λ_r, Γ_n are the values of λ and γ_n^{Cr} for $W = W_r$. For neutrons of energy less than 100 eV. the partial widths of any levels for neutron emission are almost certainly* much less than those for γ-radiation, and the variation of the latter over the small range of energies corresponding to the total level width is so small that we can replace $\gamma_{Q\beta}^{Cr}$ by its value at resonance. The cross-section for radiative neutron capture is then given by

$$\sigma(n, \gamma) = \frac{\lambda\lambda_r}{8\pi}\left(1 \pm \frac{1}{2j+1}\right)\frac{\Gamma_n\Gamma}{(W-W_r)^2+\frac{1}{4}\Gamma^2}. \qquad (8·3)$$

Here $\Gamma(= \sum_{Q\beta}\Gamma_{Q\beta})$ is the total level width, in this case due to all possible γ-transitions from the level 'r' of C.

The capture cross-section at resonance, $W = W_r$, is simply

$$\sigma_r(n, \gamma) = \lambda_r^2/2\pi(1 \pm 1/(2j+1))(\Gamma_n/\Gamma). \qquad (8·4)$$

The cross-section for resonant elastic scattering† is obtained by replacing $\gamma_{Q\beta}^{Cr}$ by γ_n^{Cr}, which leads to

$$\sigma_r(n, n) = \lambda_r^2/2\pi[1 + 1/(2j+1)](\Gamma_n/\Gamma)^2. \qquad (8·5)$$

* Except, possibly, for very light nuclei.

† The contribution to the elastic scattering not arising from resonance, i.e. the so-called 'potential scattering', and, in particular, interference between potential and resonance scattering, is here neglected.

The ratio of capture to elastic-scattering cross-section at resonance is (Γ/Γ_n). This is in agreement with the experimental observation that the cross-section for resonant scattering is much smaller than that for capture, and therefore the assumptions made about the partial widths in $(8\cdot3)$ are justified. When emission of a fast-charged particle, in addition to γ-radiation, is possible, then the total widths of a level may be much larger than W_r (depending on the barrier penetrability for charged particle emission). If this is so, then in $(8\cdot3)$ (it is still assumed that partial width for fast-particle emission is sensibly constant over the energy range involved)

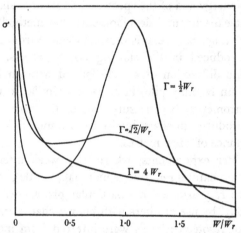

Fig. 10. Slow neutron capture. Variation of cross-section with neutron energy. (Bethe, *Rev. Mod. Phys.* 9, 119, 1937.)

the denominator can be replaced by $\tfrac{1}{4}\Gamma^2$, and then the dependence of σ on W is contained entirely in the factor λ ($\sigma \propto \lambda \propto 1/v$). This is the case for $B^{10}(n,\alpha)Li^7$ and $Li^6(n,\alpha)H^3$, in which the predominant contribution to Γ arises from α-emission. For medium nuclei the energy available for α-emission in the (n,α) reaction is about 4 MeV., for which energies the partial width for α-emission will be less than that for γ-radiation. In the interaction of slow neutrons with stable (i.e. non β-active) nuclei proton emission will be either energetically impossible or very improbable on account of the low proton energy (cf. §9).

Fig. 10 shows the theoretical variation of $\sigma(n,\gamma)$ with W for some characteristic cases [52].

8·1. Experimental methods

Nuclear reactions do not provide slow neutrons directly (due to the very small probability of emission of slow neutrons from a compound nucleus). Slow neutrons are obtained by slowing down fast neutrons produced in a nuclear process (e.g. $Be^9(\alpha, n)$, $Be^9(\gamma, n)$, $H^2(d, n)$ or fission) by nuclear collisions in a medium containing light nuclei with a small neutron-capture cross-section (e.g. H^1 as paraffin, H^2 as heavy water, carbon as graphite). In this way a continuous energy spectrum of neutrons is obtained depending primarily on the physical temperature of the medium, but also, for the low-energy and of the spectrum, on its crystalline state.*

In recent experimental developments two methods have been used for isolating a particular energy component from the complete spectrum produced in the slowing down process. These are: (i) crystalline diffraction of a well-defined neutron beam (from uranium chain reacting pile) in a manner similar to that used in X-ray spectrometry; (ii) measurement of the time of transit of neutrons (produced in sharp bursts from a modulated cyclotron) across a distance of a few metres.

In the older experiments, where only weak intensities were available (from natural radioactive material plus Be), the energy of the neutrons responsible for a particular process was measured, approximately, by use of filters with known characteristics, or the properties of resonance levels were inferred from measurements using the whole neutron spectrum. These methods have been largely superseded by the more precise techniques now available, so that they will be mentioned only briefly here.

8·2. Boron method for W_r (53)

It is clear from fig. 10 that when neutrons having a continuous energy distribution (e.g. an approximately Maxwellian distribution) pass through matter, absorption will be important for neutrons with energies in one of two ranges: (i) very low energies, so-called 'thermal' neutrons; (ii) energies near a resonance level. The cross-section for absorption of slow neutrons by boron is found experimentally to obey the '$1/v$' law accurately, as would be

---

* A neutron of energy 0·02 eV. (approximately room temperature) has a wave-length of 3×10^{-8} cm.

expected. Hence, the absorption by boron can be used to indicate the energy of slow neutrons responsible for a particular process, e.g. the energy of neutrons responsible for resonance capture in a particular element with the formation of a β-active isotope. (All elements with odd Z (except $Z = 3, 5, 7$) form β^--active isotopes on capture of a neutron, and this class also includes all elements occurring as a single stable isotope.) The β-activity is first measured with no absorber and then with a boron absorber, and then the measurements are repeated with an additional cadmium absorber to remove all 'thermal' neutrons. From these measurements the absorption in boron of both thermal and resonance neutrons, μ_{th}, μ_r, can be obtained directly, and since the cross-section for absorption in boron is proportional to $W^{-\frac{1}{2}}$, the value of the resonance energy is given by

$$W_r/W_{th} = (\mu_{th}/\mu_r)^2,\qquad (8\cdot6)$$

where W_{th} is a suitable average for the energies of thermal neutrons absorbed by boron. On the assumption of a Maxwell distribution of neutron velocities with equivalent temperature T, W_{th} equals $\frac{1}{4}\pi kT$. A correction has to be made for the small cross-section for *scattering* which does not obey the '$1/v$' law. The results obtained by this method tend to give too small a value for W_r owing to the presence of more fast neutrons than indicated by the Maxwell distribution function (54, 55).

8·3. Self-detection method for level widths

The total activity $\mathscr{A}$ produced by absorption of neutrons having an energy spectrum designated $I(W)$ is given by

$$\mathscr{A} \propto \int I(W)\,W^{\frac{1}{2}}\,\sigma(W)\,dW.\qquad (8\cdot7)$$

If the total activity is measured and $I(W)$, as known then, from $(8\cdot7)$ and $(8\cdot3)$, an estimate can be made of the quantity (Γ_n/W_r^2), provided only resonance absorption is important, i.e. thermal neutrons have been removed. This is achieved by the usual technique of absorption by cadmium. For neutrons diffusing through a slowing down medium, $I(W)$ is proportional to $W^{-\frac{1}{2}}$ for neutrons of energies greater than about 1 eV. The simplest application of $(8\cdot7)$ would then be to the interpretation of the observed activity in a *thin* detector with a known neutron energy distribution. If the detector is not sufficiently thin the energy distribution will vary with depth of penetration into the detector, since the neutron flux will be depleted most readily of those neutrons with energy near W_r. Thus the measured absorption coefficients, as indicated by the change in activity, using the same material for absorber and detector, will decrease as the thickness of absorber increases, since for thick absorbers the only neutrons left to be absorbed will be those with energy on the wings of the absorption line. When all the neutrons with energies near W_r have been absorbed, the absorption curve ceases to be exponential ($e^{-\mu x}$) and takes the form $ax^{-\frac{1}{2}}$ (x is the thickness of absorber). Analysis of the absorption curve obtained in the method of self-detector enables an estimate to be made of $\sigma_r(n, \gamma)$, i.e. of $[1 \pm 1/(2j+1)][\Gamma_n/\Gamma W_r]$ (55).

In practice the cross-section for absorption is not given directly by (8·3), since, on account of the thermal motion of the absorbing nuclei, the effective energy of incident neutrons of energy W_0 will be spread out (Doppler broadening of level) over a range of energies having the Gaussian distribution

$$N(W)\,dW = (2\pi D)^{-\frac{1}{2}} \exp\left[-(W_0 - W)^2/2D^2\right] dW, \qquad (8\cdot8)$$

where $D = 2\sqrt{\{(m/M)W_0 kT\}}$, m/M is ratio of mass of neutron to that of absorbing nucleus and T is the absolute temperature of the absorbing material.

The cross-section for neutrons of energy W_0 will be therefore

$$\sigma(W_0) = (2\pi D)^{-\frac{1}{2}} \int_{-\infty}^{+\infty} \sigma(W) \exp\left[-(W_0 - W)^2/2D^2\right] dW. \quad (8\cdot9)$$

If $D \gg \Gamma$, then the variation of σ near W_r is primarily governed by the Doppler broadening and the value of σ_r is reduced by a factor Γ/D, but for energies sufficiently far removed from W_r the Doppler effect is unimportant, since the Doppler factor falls off exponentially whereas in the expression (8·3) $\sigma(n, \gamma)$ falls off quadratically with $(W - W_r)$. As far as the actual measurements are concerned, absorption with self-detection using thick absorbers gives a result independent of Doppler broadening, since only the wings of the resonance peak are important. The absorption as measured with *thin* absorbers will be reduced by a factor Γ/D (if $\Gamma \ll D$).

Analysis of the absorption curve using self-detection has been useful in checking whether or not the absorption is due to a single resonance peak; the shape of the absorption curve is dependent on the number (and position) of the resonance levels.

8·4. Mechanical 'velocity selector' (56)

An intermittent source of neutrons is created by opening and closing rapidly a shutter consisting of a pair of disks, one stationary and the other rotating, fitted with cadmium sectors, and placed in front of an ordinary source of slow neutrons. A second similar shutter is placed on a common axis, some distance away, in front of the neutron detector. This whole arrangement acts as a slow neutron velocity filter. With fifty sectors on each disk and the two shutters placed 50 cm. apart and rotating at a maximum speed of 5000 r.p.m., the most energetic neutrons that can be 'filtered', have an energy of about 0·25 eV. Mechanical difficulties, together with limited intensities that can be produced, restrict the range of this technique to the region of thermal energies, and even for this region the method has been largely superseded by the more powerful (and more elaborate) electrical and diffraction methods.

8·5. Electrical velocity selector

In this technique, which was first used by Alvarez and developed by Baker and Bacher and Rainwater and Havens and others (57), the ion source of a cyclotron is modulated so as to produce an intermittent beam of deuterons. The deuterons bombard a beryllium target producing short bursts of neutrons which are slowed down in the usual way, by passage through a few centimetres of paraffin wax. The slow neutrons are detected by an ionization chamber filled with BF_3 (ionization produced by α-particles from the reaction $B^{10}(n, \alpha) Li^7$), which is placed at a distance of several metres from the neutron source. Ionization pulses are amplified and recorded in different channels corresponding to different times of transit of the neutron from the source to the detecting chamber, i.e. different energies. In some recent experiments the pulse of deuterons was variable between 10 and 1000 μsec. duration (depending on the energy range being investigated), and was repeated either 1000 or 100 times a second. The distance between source and detector was 5·4 m., and neutrons were detected concurrently in sixteen separate channels each corresponding to a different neutron energy. Thermal neutrons from the source were removed by cadmium or boron absorbers. The actual transit time of the neutron is not given directly by the time interval between the current pulse modulating the cyclotron and the voltage pulse which opens the particular recording channel for a short interval (5–10 μsec.). Corrections, amounting to about 25 μsec. in a typical arrangement, have to be made (and determined experimentally) for the finite time spent by the deuterons in their orbits in the cyclotron, for the slowing down time of the fast neutrons in the paraffin (especially important at small energies, of order 1 eV. or less) and for the finite time of collection of ions in the BF_3 detecting chamber. The energy *resolution* is determined by the finite time intervals during which the source and detecting channels are switched on, the finite length of the BF_3 chamber (about 10 cm.) and the duration of the amplified voltage pulse produced by a single detected neutron.

The spread in energies δW, for a single detecting channel, corresponding to a spread in transit time δt, is*

$$|\,\delta W\,| = 2W(\delta t/t), \qquad (8·10)$$

and for the finite spread δL in transit distance L

$$|\,\delta W\,| = 2W(\delta L/L). \qquad (8·11)$$

For energies of 10 eV. or more (with $L = 5·4$ m., $W = 10$ eV., corresponding to $t = 124 \times 10^{-6}$ sec.) $\delta t/t \gg \delta L/L$, so that even with the shortest neutron bursts (10^{-5} sec.) the resolution is governed by the finite duration of the timing pulses.

Measurement of the *shape* of a neutron resonance level requires an energy resolution of the order 0·03 eV. or less (typical level widths, Γ_r, are about 0·1 eV.), which is only obtainable with the arrangement described above, when the neutron energy is ∼ 1 eV. or less. If the *spacing* of the resonance levels in the spectrum is about 10 eV. (a reasonable average), then observations of individual levels will be reliable only for energies less than about 25 eV.

The experimental procedure consists of examining the ratios of the numbers of neutrons detected in each recording channel

---

* The effective shape of the energy spectrum admitted by a single recording channel is triangular with a base length determined by the factors limiting resolution.

(corresponding to a series of different adjacent energy intervals) with and without a sheet of material, containing the nuclei under investigation, in front of the detecting chamber. These ratios give a measure of the total cross-section, i.e. the sum of the scattering and capture cross-sections, at a series of different neutron energies. If the absorber is thin, i.e. the fraction of the neutrons absorbed or scattered in any of the energy intervals (including $W \approx W_r$) is very much less than unity, then the measurement of the absorption, $\delta I/I$, gives directly a measure of the total cross-section, averaged over the particular energy interval

$$(\delta I/I)_{Av} = \overline{\sigma(W)} n \delta x,$$

where n is the number of nuclei per c.c., δx is the thickness of the absorber, and the average is over the energy interval. If σ is regarded as constant over the energy interval, then

$$I(x) = I_0 e^{-\sigma(W)x}. \tag{8.12}$$

If, however, the value of σ varies appreciably over the individual energy intervals (which possibility is most serious for W near W_r), then the absorption will not follow a simple exponential law, and a more detailed analysis is necessary. This usually involves a trial and error procedure to obtain the value of $\sigma(W)$ (57).

For maximum statistical accuracy $I(x)/I(0)$ should be about 0·1, but except for resonance levels occurring at very low energy (less than 1·0 eV.), where very high resolution is available (cf. (8.10)), such large absorption ratios cannot be used in practice if (8·12) is to be a reasonable first approximation. For different parts of the energy spectrum corresponding to different values of $\bar{\sigma}$, different absorber thicknesses are used so as to keep the value of $I(x)/I(0)$ about 0·4; and, in addition, the various time intervals are adjusted to two or three different scales corresponding to different energy ranges.

When the shape of the resonance line is fully resolved ($E \lesssim 1$ eV.) the absorption measurements give directly W_r, Γ (total level width) and σ_r; and assuming $\Gamma_n \ll \Gamma$, the quantity $\Gamma_n(1 \pm 1/(2j+1))$ is obtained. If the shape is only partly resolved ($\Gamma \sim \delta W$) then, in addition to W_r, only an *upper limit* to Γ and a *lower limit* to σ_r are obtained directly, and those values of Γ and σ_r which give the best fit with the experimental results are adopted. Examples of such measurements by Rainwater and Havens *et al.* (57) are shown in fig. 11.

For energies greater than about 5 eV., $\delta W \gg \Gamma$ (if Γ is assumed to be approximately 1 eV. or less), and then the maximum absorption observed corresponds to the *average* value of the absorption over an energy range δW centred on W_r. The measurements then provide a value for the quantity $\sigma_r \Gamma^2$, the so-called 'strength' of the resonance level.

For neutrons of thermal energies ($W < 0.1$ eV.) the total cross-section for nearly all nuclei is found to vary with energy according to

$$\sigma(\text{total}) = \sigma(\text{capture}) + \sigma(\text{scatt.}) = AW^{-\frac{1}{2}} + B. \qquad (8.13)$$

The first term arises from capture and the second from elastic scattering (cf. (8.3), (8.5)). Measurement of the absorption of thermal neutrons provides an estimate of B which is then applied as a correction term at all energies to obtain σ(capture). At very low energies where the wave-length of the neutron ($h/\sqrt{(2mW)}$) is comparable with the crystal-lattice spacing of the absorbing material, the interaction of neutrons can no longer be considered as occurring independently with individual nuclei. Interference between the elastically scattered neutrons from individual nuclei is detectable, and the simple expression (8.12) for absorption is no longer valid. Processes other than elastic scattering are not coherent, so that interference can be ignored.

Apart from the low resolution at large energies the other major limitation of the 'time-of-flight' technique is the ambiguity in the interpretation of the results when the absorbing element contains more than one isotope.* If only a single resonance is observed, it is possible to produce the same resonant absorption process by irradiating the element in question with a flux of slow neutrons from which thermal neutrons have been removed (cadmium absorption), and then measuring the characteristic decay constant of the product nucleus (if β-active). If the same

---

* Not only does this lead to ambiguity in attributing the resonance to a level in a particular nucleus, but also, since the number of nuclei per unit volume, n, is not known with certainty, σ cannot be calculated from (8.12). In the literature it is usual to state the value of σ obtained by using for n the *total* number of nuclei of all isotopic masses. This gives too small a value for σ which can be corrected only if the isotope responsible and its relative abundance is known, except in cases where all isotopes are present roughly equal abundance.

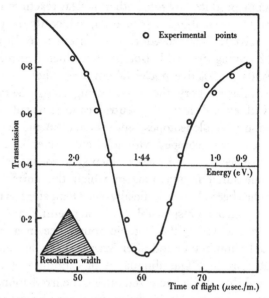

Fig. 11. (a) Slow neutron transmission of indium (0·193 g./cm.²). Curve: Breit-Wigner dispersion formula with $W_r = 1·44$ eV., $\Gamma = 0·09$ eV. (Havens and Rainwater, *Phys. Rev.* **70**, 160, 1946.)

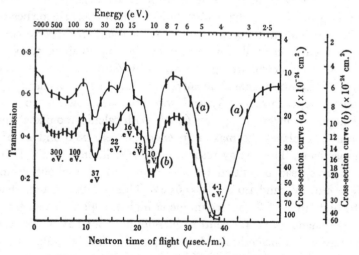

Fig. 11. (b) Slow neutron transmission of tantalum. (a) for 9·98 g./cm.², (b) for 22·44 g./cm.². (Havens, Wu, Rainwater and Meaker, *Phys. Rev.* **71**, 168, 1947.)

activity can be produced by some other nuclear reactions it is often possible, from the combined information, to identify the particular nucleus involved. This procedure is often effective for elements with odd Z having two stable isotopes for which neutron capture always results in β-active nuclei (about 40 % the elements with odd Z are in this category, the others having a single isotope). For elements with even Z, however, neutron capture usually results in the formation of stable isotopes, since these elements have several naturally occurring isotopes with adjacent values of A. Other methods have been used for these elements. Resonance absorption has been measured in materials in which the normal isotopic abundance has been altered artificially (59). Dempster has been able to identify the element responsible for strong neutron absorption in cadmium by measuring the isotopic abundance in a sample of ordinary cadmium before and after irradiation by an intense flux of thermal neutrons. The abundance of cadmium of mass 114 increased and that of 113 decreased during the irradiation, proving that the latter is responsible for the strong absorption (60).

8·6. Neutron levels by crystal spectrometry

From the very intense slow-neutron fluxes available in chain-reacting piles it is possible to obtain highly collimated neutron beams of sufficient intensity for the methods familiar in X-ray spectroscopy to be used successfully for neutron energies up to about 50 eV.

The experimental arrangement used by Zinn (61), Sturm (62) and others is shown diagrammatically in fig. 12. Using crystals of LiF (100 plane, separation 2Å), and grazing angles down to 30′, the maximum neutron energy was 65 eV. With calcite (100 plane) and mica (001 plane) the maximum energies were 28·0 and 2·5 eV. respectively. The energy resolution $\delta W/W$, limited by the finite angular spread permitted by the collimators, was about $\frac{1}{30}$ for $W = 0\cdot01$ eV., and $1\cdot0$ for $W = 65$ eV. This is rather poorer than the best 'time-of-flight' technique at high energies. The shape of the resonance peaks for neutron energies of about 1 eV. or less can be resolved sufficiently faithfully to permit the level width, Γ, to be determined directly. (Performance in this respect about the same as best 'time-of-flight' measurements.) In the experiments that have been made so far the total absorption has been measured as a function of energy using a BF_3 chamber as detector. Since the crystal spectrometer actually separates out spatially different energy components of the neutron spectrum, it is possible in

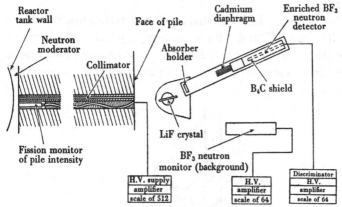

Fig. 12. Neutron spectrometer. (Sturm, *Phys. Rev.* **71**, 759, 1947.)

principle (and in practice if intensity permits!) to measure the characteristic decay time corresponding to a particular resonance level, in cases where capture leads to a β-active nucleus. This would be of value in identifying the particular nucleus concerned.

Results obtained by Sturm with iridium ($Z = 77$, $A = 191$, 193) absorbers are shown in fig. 13.

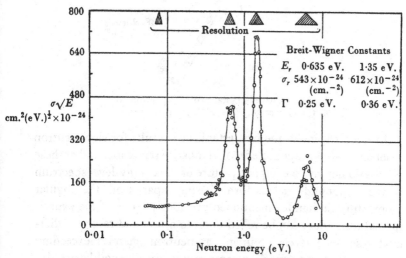

Fig. 13. Total cross-section iridium. (Sturm, *Phys. Rev.* **71**, 771, 1947.)

Measurements have also been made recently (63) with a curved-crystal spectrometer using a convergent beam of slow neutrons of angular width 9′. The general arrangement is shown in Fig. 14.

The resolution $\delta W/W$, obtained with this technique, was about $\frac{1}{50}$ for $W = 0.05$ eV. and $\frac{1}{8}$ for $W = 2.0$ eV., which again is rather poorer than the best 'time-of-flight' experiments. The results obtained by this method are similar to those using the plane-crystal spectrometer.

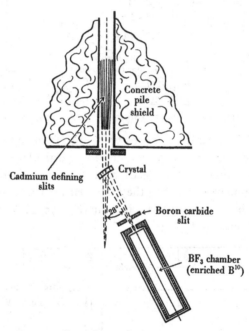

Fig. 14. Curved-crystal neutron spectrometer.
(Sawyer *et al.*, *Phys. Rev.* **72**, 110, 1947.)

In table 1 some of the best established results for slow-neutron resonance levels are collected (51). It must be remembered that these results do not give a complete picture of the energy-level spectrum of the compound nucleus formed, since apart from the angular momentum limitations mentioned previously, none of the experimental techniques can be relied upon to detect and resolve individual resonance levels occurring at neutron energies exceeding about 30 eV. For comparison some proton-capture widths are also

TABLE 1. *Table of neutron and proton capture resonance levels*

Neutron

Element (Z)	Isotopes	W_R (eV.)	Γ (eV.)	$\sigma_r \Gamma^2$	σ_r ($\times 10^{-24}$ cm.²)	Γ_n (eV.)	Γ_n (1 MeV.) (eV.)	Experimental methods*
Cd (48)	106, 108, 110, 111, 112, 113, 114, 116	0·178	0·115	—	>7,000	$>10^{-4}$	~1	T, C, P
In (49)	113, 115	1·44	0·09	210	2,600	$2·7\times10^{-3}$	2·2	T
Eu (63)	151, 153	0·465	0·2	—	11,400	$0·85\times10^{-3}$	1·25	T, P
Dy (66)	158, 160, 161–164	−1·01	0·1	—	79,000	—	—	P
Au (79)	197	5·4	<2	600	>200	$2·2\times10^{-3}$	~1	T, P

General reference: Goldsmith, Ibser and Feld, *Rev. Mod. Phys.* **19**, 259, 1947.

* T = time of flight, C = curved-crystal spectrometer, P = plane-crystal spectrometer.

Proton

Target nucleus	Compound nucleus	W_R (keV.)	Γ (keV.)	$g\Gamma_\gamma$ (eV.)	Γ_p (1 MeV.)† (no barrier) (keV.)
Li$^7_3$	Be^8	439	12	9	71
Be$^9_4$	B^{10}	988	94	12·5	235
Be$^9_4$	B^{10}	1077	4	0·8	33
Al$^{27}_{13}$	Si^{28}	1368	9·8	—	~200

General reference: Fowler, Lauritsen and Lauritsen, *Rev. Mod. Phys.* **20**, 236, 1948.

† Allowance made for barrier penetrability (Christy and Latter, *Rev. Mod. Phys.* **20**, 185, 1948).

presented. Partial widths for neutron and proton emission are normalized to 1 MeV. by assuming $\Gamma \propto E^{\frac{1}{2}}$ and allowing for barrier penetrability in the case of protons (cf. §9·1).

8·7. Slow neutron scattering (64)

According to (8·1) the cross-section for elastic scattering should also exhibit resonance properties. The situation is complicated in this case by the presence of 'potential scattering', i.e. scattering due to the interaction of nucleus and neutron without formation of a compound nucleus in a definite state. The amplitudes of the scattered waves due to the two processes must be added with correct regard to their relative phases. As long as $\Gamma \gg \Gamma_n$ (and this probably holds for neutron energies up to 10^3 eV. in most nuclei), the cross-section for resonant scattering will be smaller than for capture in the ratio Γ_n/Γ. Many investigations have been made of the scattering of slow neutrons, but none have included a study of the variation with energy from which the properties of individual resonance levels can be deduced.

8·8. Fast neutron processes

Neutrons with energies in the region 10^3–10^5 eV. are difficult to study experimentally, and very little is known directly about their interaction with nuclei. At energies greater than 10^5 eV. (i.e. 'fast' neutrons) we should expect elastic scattering rather than capture to be the predominant process (unless the emission of a fast charged particle is a more probable process). Neutrons with energy in the range 0·1–10 MeV. cannot, with present techniques, be produced with adequate intensity and energy spread of less than about 10^3 eV., so that observation of discrete levels will be limited to fairly light nuclei with large level spacing.

In many cases the cross-section for nuclear reactions or scattering, with neutrons having an energy of 1 or 2 MeV., show pronounced maxima at particular energies, but only in a few cases can individual levels be recognized.

Using a helium-filled cloud chamber and a continuous neutron spectrum (0·6–2·0 MeV.), Staub and Tatel (65) were able to establish the presence of two close levels in He^5 as indicated by maxima in the elastic scattering cross-section, $He(n, n)$. These levels of He^5 have an energy in excess of a free neutron and He⁴ of about 0·9 MeV., widths of approximately 0·4 MeV., and the level separation is 0·25 ± 0·1 MeV. No disintegration is possible in this case, and since there is no stable He⁵ nucleus radiative capture is also impossible.

An example in which the usual processes are possible and discrete resonance levels are resolved is provided by the variation of the total cross-section for fast neutrons in Al²⁷, as measured by

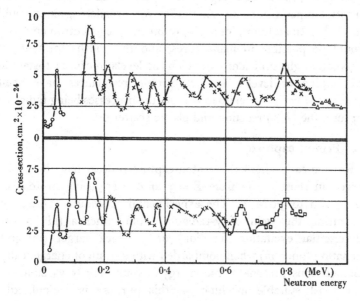

Fig. 15. Total neutron cross-section for Al²⁷. (Neutrons from Li⁷ (p, n). Circles represent measurements taken at 115° with respect to proton beam. All other points for neutrons in forward direction. Seagondollar and Barschall, *Phys. Rev.* **72**, 442, 1947.)

Seagondollar and Barschall (66). Neutrons of variable energy were obtained by bombarding thin Li targets with homogeneous protons, (Li⁷ (p, n) Be⁷). Their results are shown in fig. 15.

The widths of most of the resonance levels in this case are probably due to limited experimental resolution, so that the experimental cross-sections at resonance are lower limits only. The widths of the lowest observed level (at 0·155 MeV.) is less than 16 × 10³ eV.

9. RESONANCE LEVELS IN PROTON REACTIONS

It is not likely that discrete resonance levels can be observed in proton reactions for nuclei with $Z > 30$. For such nuclei the proton energy necessary for appreciable yields would be ~ 3 MeV. or more, corresponding to an excitation energy of the compound nucleus of ~ 10 MeV. or more, at which energies the level spacing is probably less than the energy resolution experimentally attainable. (For protons of energy 5 MeV., a nucleus of mass 100 and a target temperature of 300° A., the Doppler broadening alone would be 70 eV.) In addition, for such proton energies, neutron emission would be possible in many cases, and the widths for neutron emission might sufficiently broaden the levels to cause substantial overlapping of levels. In fact, proton resonances have been observed only for nuclei up to $Z = 20$, the processes being radiative capture, the (p, α) reaction and elastic scattering.

9·1. Proton capture

Resonance levels in the proton-capture process have been observed in most nuclei from $Z = 3$ to $Z = 17$, using protons of energy up to 3·0 MeV. The experimental procedure consists of measuring the yield of γ-radiation from either a thin target (differential excitation function) or a thick target (integral excitation function) when bombarded with a beam of energetically homogeneous protons whose energy is continuously variable [67]. The most suitable machine for this purpose is undoubtedly the electrostatic generator which provides a constant but readily variable accelerating potential. Various auxiliary devices have been made to improve the homogeneity of the proton beam (usually at the expense of intensity). Amongst these may be mentioned: (1) energy focusing of a parallel beam by bending through 90° in a magnetic field [68]; (ii) similar magnetic resolution, combined with an automatic arrangement for correcting the accelerating potential when the position of the resolved beam indicates a departure of the energy from some present value [69]; (iii) an arrangement for rejecting all observations made when the voltage, as measured by a high-resistance potentiometer, differs from a pre-set value by more than a small amount [70]; (iv) energy focusing in a radial electrostatic field (71). In the best of these arrangements, the spread in proton energy is reduced to less than 1 part in 1000.

If the total level width exceeds the experimental energy resolution, then E_r and Γ can be obtained directly, and the absolute yield of γ-radiation at resonance provides an estimate of σ_r, provided that the number of nuclei/c.c. in the target is known and the yield is measured for a sufficiently representative range of angles for the total yield to be estimated. Since the partial width for γ-radiation, Γ_γ, is not likely to exceed a few volts, the total width will only exceed the experimental resolution (say 10^3 eV.) if the proton width is sufficiently large (and approximately equal to the total width). The expression for the cross-section then becomes

$$\sigma(p, \gamma) = \lambda^2/4\pi g[\Gamma_\gamma \Gamma/\{(W - W_r)^2 + \tfrac{1}{4}\Gamma^2\}], \qquad (9\cdot1)$$

where g is a spin-statistical factor of the order unity and depends on the angular momentum of the compound state in addition to the intrinsic angular momenta of the nuclei involved. The cross-section at resonance is

$$\sigma_r(p\gamma) = (\lambda_r^2/\pi)g(\Gamma_\gamma/\Gamma), \qquad (9\cdot2)$$

so that $g\Gamma_\gamma$ can be determined when Γ, W_r and the absolute yield are known. If the level width is much less than the experimental resolution then only the cross-section averaged over the level is measurable. This is usually obtained by measuring the γ-radiation yield from a *thick* target using a proton beam energy exceeding W_r by at least the energy spread of the beam. Under these conditions the yield of γ-radiation will be independent of W (until a higher resonance level is reached) and is equal to

$$Y(\gamma) = n/\kappa \int_r \sigma(W)\,dW. \qquad (9\cdot3)$$

Here Y is the yield of γ-quanta per proton (assuming a single γ-transition following each proton capture), n the number of capturing nuclei per c.c. in the target, $\kappa = dW/dx$, i.e. rate of loss of energy experienced by proton in penetrating the target. The integration is carried out over all energies near W_r for which $\sigma(W)$ is appreciable. Two limiting cases give particularly simple results:

(i) $\Gamma_p(\approx \Gamma) \gg \Gamma_\gamma$, then using (9·1), (9·3) becomes

$$Y(\gamma) = (n/2\kappa)g(h^2/2mW_r)\,\Gamma_\gamma, \qquad (9\cdot4)$$

which is independent of Γ_p! (m is the reduced mass of the proton).

(ii) $\Gamma_\gamma(\approx \Gamma) \gg \Gamma_p$ (this may be true for very small proton energies). Then

$$Y(\gamma) = (n/2\kappa)g(h^2/2mW_r)\,\Gamma_p, \qquad (9\cdot5)$$

which is independent of Γ_γ!

In intermediate cases, we have the result:

$$Y(\gamma) = (n/2\kappa)g(h^2/2mW_r)(\Gamma_\gamma\Gamma_p/\Gamma), \qquad (9\cdot6)$$

which is valid even if other competitive processes occur.

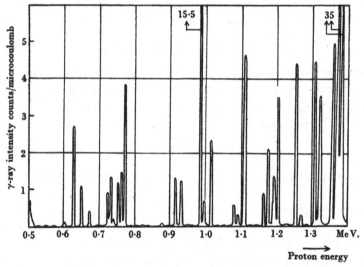

Fig. 16. Al (p, γ) yield curve. (Broström, Huus and Tangen, *Phys. Rev.* **71**, 663, 1947.)

We see, then, that when Γ is too small to be measured directly, measurement of the absolute 'integrated' yield of γ-quanta together with an estimate of the rate of loss of energy of protons in the target material provides a measure of the smaller of the quantities Γ_γ and Γ_p, provided other competing processes are absent.

Fig. 16 shows the results of recent investigations of the $Al^{27}(p, \gamma)$ process by Broström *et al.* (70), which clearly illustrate the high resolution attained.

From the measured yield, $g\Gamma_p$ is estimated at about 1 eV. for protons of 0·6 MeV. energy, and for higher energies $g\Gamma_\gamma$ appears to range from 0·15 to 17·0 eV. The latter values are rather large for partial γ-widths and probably correspond to several possible γ-transitions.* A full discussion of the technique of measurement of the absolute γ-yield and measurement of the level widths is given in a recent review by Fowler, Lauritsen and Lauritsen (72).

An interesting and well-known instance of a sharp resonance level for the (p, γ) process in which the compound nucleus has sufficient energy to emit α-particles with energy of several MeV. occurs for Li^7 with protons of 0·44 MeV. energy. A simple explanation of non-emission of α-particles (the yield of α-particles shows no resonance effects at this energy), which would greatly broaden the level of the compound nucleus Be^8, is provided by the assumption that the state of Li^7 formed has either odd angular momentum or odd parity (or both) and cannot therefore break up into two α-particles. Since α-particles obey Bose statistics and have no intrinsic angular momentum, the wave function describing two α-particles must have even parity and correspond to even total angular momentum of the system. The total γ-ray yield and level width in this case have been measured directly by several investigators (72), (73). Γ is about 10^4 eV. and Γ_γ about 20 eV. (see table 1, p. 69).

9·2. (p, α) reactions

Discrete resonance levels have been observed for the (p, α) reaction in a few light nuclei; the examples which have been studied most thoroughly are $B^{11}(p, \alpha) Be^8$ (127) and $F^{19}(p, \alpha) O^{16}$ (75). In the former case a sharp resonance level is obtained in the yield of *high-energy* α-particles ($W > 6$ MeV.) at a proton energy of 0·16 MeV. The total level width is less than 5×10^3 eV., the limit being set by experimental resolution. A sharp maximum in the yield of γ-radiation (proton capture) is also observed at the same energy, indicating competitive emission of γ-radiation and energetic α-particles from the same level of C^{12} whose excitation

---

* The emission of α-particles, $Al^{27}(p, \alpha) Mg^{24}$, in competition with protons and γ-radiation, would make the values of $g\Gamma_\gamma$ given here *lower* limits.

energy (as deduced from nuclear masses) is 16·16 MeV. The small level width is attributed to some selection rule which reduces the probability of α-emission by a factor of 100 or more.*

In the reaction $F^{19}(p, \alpha) O^{16}$ the sharp resonance levels correspond to maxima in the yield of *low-energy* α-particles[68] ($W \sim 2$ MeV.), O^{16} being produced in an excited state with excitation energy of 6 MeV. (The resonance levels are usually investigated by studying the γ-radiation in the de-excitation of O^{16}.) To explain the sharpness of the resonance levels in this case it is usual to assume, in agreement with observation, that emission of α-particles of maximum energy (about 8 MeV.) corresponding to the production of O^{16} in the ground state is not possible for these states of Ne^{20} (the compound nucleus) on account of their symmetry properties. The level width for the emission of low-energy α-particles and re-emission of protons is reduced by the effect of the Coulomb barrier. The experimental results indicate a level spacing in Ne^{20} of about 0·1 MeV. at an excitation energy in the region of 14 MeV. (corresponding to a proton energy of about 1 MeV.) and level widths ranging from 10^3 to 2×10^4 eV. Competitive γ-emission has only been observed for one of these levels[74], indicating that the partial width for γ-emission is much smaller than that for α-emission. The latter probably determines the total width for proton energies less than 1 MeV.

Pronounced maxima, at energies differing from the resonance levels discussed above, are also observed in the yield of energetic α-particles ($W \approx 8$ MeV.), but the levels involved do not appear to be fully resolved. These levels also appear to be associated with the emission of less energetic α-particles leaving O^{16} in an excited state different from that responsible for γ-radiation[75]. The excited states of O^{16} in this case make the transition to the ground state by the emission of electron pairs. The identification of the resonance levels responsible for pair production and for emission of energetic α-particles cannot be made without a knowledge of the cross-section of these processes at all angles (cf. §7·1).

---

* Discrete resonance levels have also been found for the (p, α) reactions in heavier nuclei. (Baxter and Freeman, *Nature*, **162**, 696 (1948).)

10. RESONANCE LEVELS IN α-PARTICLE REACTIONS

In most of the early work using α-particles of limited energy from natural radioactive sources only light elements could be disintegrated, and for these, the level spacing is sufficiently large for resonance levels to be resolved, or at least detected. The experiments were made with necessarily poor resolution (o·1 MeV. or more) and low intensity, so that the degree of resolution of individual levels is doubtful in most cases. Two types of reaction, (α, p) and (α, n), have been extensively studied, the same resonance levels being frequently observed for both reactions. The variation in cross-section with α-particle energy is determined by measuring the variation in the yield of protons or neutrons or in the amount of β-active material produced. (The product nucleus is always stable for (α, p) reactions and always β-active for (α, n) reactions if the bombarded nucleus has odd Z and $Z > 5$.) Measurements of the cross-sections of the (α, p) and (α, n) reactions in Al^{27} are amongst the best that have been made [76]. In this case six resonance levels in the (α, p) reaction have been observed for α-particle energies between 4·0 and 6·6 MeV. Similar levels have been found for the reaction $Al^{27}(\alpha, n)P^{30}$, the energy values agreeing to within about o·1 MeV. These levels correspond to excited states of P^{31}, the compound nucleus, with excitation energies between 12·8 and 15·1 MeV. The large level spacing observed (about o·4 MeV.) is surprising for such large excitation energies in a nucleus of this size (compare, for example, the level spacings in (p, γ) reactions), indicating either that only the more pronounced levels were detected or that individual levels were not resolved. All that can be said about the level widths is that they were not noticeably larger than the experimental resolution (o·1 MeV.). For the higher levels the neutrons are emitted with an energy of approximately 3 MeV., so that if we assume the neutron width to be proportional to $W^{\frac{1}{2}}$, the width for neutron emission with energy 1 MeV. would be less than o·06 MeV.

11. RESONANCE LEVELS IN DEUTERON REACTIONS

The excitation energy of the compound nucleus formed in deuteron reactions is so large (and the emission of an energetic neutron or proton is always possible) that it is rather surprising to find sharp resonance levels in reactions of this type. However, sharp maxima in the variation of cross-section with energy have been observed in a few cases (77), notably the reaction $C^{12}(d, n) N^{13}$, for which several maxima, representing partly overlapping levels, have been found in the deuteron energy range 0·9–1·9 MeV. The narrowest maximum, at a deuteron energy of 1·33 MeV., has a width (experimental?) of about 2×10^4 eV. If the Oppenheimer-Phillips process predominates (v. § 2·5), maxima in (d, p), or (d, n), cross-sections may relate to a high probability of capture of neutrons (or protons) from the bombarding deuterons, analogous to resonant neutron (proton) capture rather than excitation of a level of the whole compound nucleus.

12. RESONANCE SCATTERING OF CHARGED PARTICLES

In estimating the cross-section for elastic scattering of charged particles it is necessary, in a full analysis, to consider three factors: (i) scattering in which a compound nucleus is formed, (ii) potential scattering by the specifically nuclear field of force, and (iii) the modification of the incident beam of particles by the Coulomb field before reaching the nuclear surface ('Rutherford scattering'). Since Coulomb forces are of the long-range type ($V(r) \sim 1/r$), the scattering cross-section due to them is strongly dependent on angle. It is therefore necessary to consider, instead of the total scattering cross-section, the differential cross-section $\sigma(\theta)$ for scattering into unit solid angle at a particular angle θ to the incident particles (all angles reduced to centre-of-gravity co-ordinate system).

For scattering by a pure Coulomb potential, the cross-section is given by the Rutherford formula

$$\sigma_c(\theta) = (zZe^2/4W)^2 \operatorname{cosec}^4(\tfrac{1}{2}\theta), \qquad (12 \cdot 1)$$

where ze, Ze are the charges of the scattered and scattering nuclei.

An estimate of the order of magnitude of the resonance scattering is obtained from (7·2) by making the assumption that no other process has comparable probability. Then for the average differential cross-section at resonance we can write

$$\sigma_r(\theta) \sim g\hbar^2/2mW, \qquad (12\cdot2)$$

where m is the reduced mass of the scattered particle and g a statistical factor of order unity. Considering the two processes as independent (in fact, interference between scattering of both types occurs), we can get an estimate of the conditions under which resonance scattering will be detectable in the presence of Coulomb scattering, viz.

$$\overline{\sigma_r(\theta)}/\sigma_c(\theta) \sim \sin^4(\tfrac{1}{2}\theta)g(8W/m)\,\hbar^2/(zZe^2)^2$$
$$\sim 3\cdot5 \times 10^3 g \sin^4(\tfrac{1}{2}\theta)(W/Az^2Z^2), \qquad (12\cdot3)$$

where W is in MeV. and A is atomic mass units. In practice the ratio of resonance and Coulomb scattering may be reduced below the real value by a factor $\Gamma/\delta W$ if the level width Γ is less than the energy spread δW of the scattered particles. It is clear from $(12\cdot3)$ that resonance scattering will be most readily detectable for small Z, large scattering angles $(\theta \sim 180°)$ and with energies, W, for which the level width (strongly dependent on W due to barrier penetrability) is comparable with the experimental resolution and yet not so large that the discrete level structure disappears. It is also necessary that there should be no competitive process which is much more probable than elastic scattering.

In the theoretical treatment of the problem the nuclear potential scattering is usually neglected to avoid excessive complication. Bethe[78] has given an expression for the scattering cross-section arising from a single resonance level in terms of the ordinary Rutherford scattering cross-section for the same nucleus and energy:

$$\sigma(\theta)/\sigma_c(\theta) = 1 + (\rho^2 + 2\rho \sin \xi + 2\rho x \cos \xi)/(1+x^2);$$
$$x = 2(W-W_r)/\Gamma; \quad e^{i\xi} = e^{i\alpha \log \sin^2(\tfrac{1}{2}\theta)}\left[\frac{(1+i\alpha)^2 \dots (J+i\alpha)^2}{(1+\alpha^2)\dots(J^2+\alpha^2)}\right];$$
$$\alpha = (zZe^2/\hbar)\sqrt{(m/2W)}; \quad \rho = [2(2J+1)/\alpha]\sin^2(\tfrac{1}{2}\theta)P_J(\theta)(\Gamma_s^r/\Gamma).$$
$$\dots(12\cdot4)$$

Here Γ_s^r is the partial width of the level, r, for elastic scattering, Γ the total width and J is the total angular momentum of the resonance level of the compound nucleus. In this expression it is assumed that both scattered particle and scattering nucleus have zero intrinsic angular momenta, and only a single resonance level of the compound nucleus is involved (i.e. no overlapping). The first two terms in the expression, i.e. 1 and $\rho^2/(1+x^2)$, represent the separate effects of Coulomb and resonance scattering, and the remaining two terms arise from interference effects between them. Due to this latter contribution the variation of $\sigma(\theta)$ with W does not exhibit a simple maximum for a single resonance level, but $\sigma(\theta)$ passes through *maximum* and *minimum* values in an energy range of magnitude Γ about the mean value W_r. From $(12\cdot4)$ it can be seen that for values of θ such that $P_J(\theta) = 0$, $\sigma = \sigma_c$ for all energies.* This result can be used to find J for a particular level, and hence also ξ. In fitting the experimental observations to particular values of Γ and W_r it is necessary to take into account the finite energy resolution in the experiment. A systematic method of allowing for this factor has been suggested by Rose [79].

Apart from the investigation of the collisions of very simple systems (e.g. p, n, d) where the formation and properties of a compound nucleus do not provide appropriate descriptions of the processes, most of the experimental investigations have been made with natural α-particle sources (Ra C', Th C') and with rather poor energy and angular resolutions ($\delta W \sim 0\cdot2$ MeV., $\delta\theta \sim 10°$ [80]). Of the elements investigated, O^{16}, C^{12} both possess zero spin, so that the theoretical expression $(12\cdot4)$ is applicable. Evidence of resonance scattering is obtained in both cases, but the assignment of values to Γ and W_r cannot be made precisely. Other elements that have been investigated (He^4, F^{19}, N^{14}) also show resonance scattering [80] but require a more elaborate theoretical treatment for their interpretation.

For He^4, the complication arises from the identity of the scattering and scattered particles so that the wave function must be symmetrical in both particles (α-particles have Bose statistics). Thus only those states of Be^8 can be formed for which $J = 0, 2, 4, \ldots$. A detailed analysis of the experimental results [81] in this case has been made by Wheeler [82], who concludes that the observed scattering can be explained by the presence of a resonance level in

---

* Also notice that for odd values of J, the minimum in σ/σ_c occurs at lower energies than the maximum and conversely for even J [79].

Be^8 with excitation energy of about 2·8 MeV. (i.e. corresponding to an α-particle energy of 5·6 MeV.) with width of 0·8 MeV. and zero angular momentum and possibly another wider level at 4–5 MeV. Evidence for the level at 2·8 MeV. is found in several nuclear reactions (e.g. B^{11} (p,α) Be^8; Li^7 (d,n) Be^8; Li$^8 \xrightarrow{\beta}$ Be^8). It is unstable against dissociation into two α-particles, and the width indicates a lifetime for this process of 10^{-21} sec. Fig. 17 shows some experimental results for the scattering of α-particles by helium,

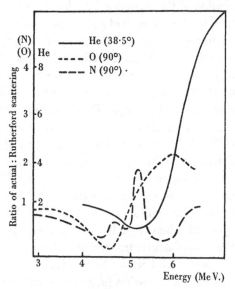

Fig. 17. Scattering of α-particles.
(Devons, *Proc. Roy. Soc.* A, **172**, 127, 559, 1939.)

oxygen and nitrogen (scattering angles in laboratory co-ordinates 38°, 90° and 90° respectively).

Recently experimental observations have been made of the resonance scattering of protons associated with some of the resonance proton-capture levels observed in light nuclei. Creutz[83] has measured the scattering of protons from a thick Li target. He observed an anomaly in the scattering cross-section in the neighbourhood of 0·45 MeV. proton energy, which indicated a resonance level with a width of about 4 × 10^4 eV. This is somewhat larger than the width deduced from proton-capture measurements (cf. §9·1), but the discrepancy may well be due to experimental inaccuracies and the neglect of interference between Coulomb and resonance scattering in the analysis of the observations. Anomalous scattering of protons by Be^9, corresponding to observed capture resonances at proton energies of 0·97 and 1·06 MeV., has been measured by Rubin [84].

13. VIRTUAL LEVELS IN MANY-PARTICLE REACTIONS

When the excitation of the compound nucleus formed in a nuclear reaction is sufficient, the emission of a particular particle group may correspond to production of a residual nucleus with excitation energy greater than the maximum for bound states. The residual nucleus can then emit a further particle. The second process may be much slower than the first, so that it is legitimate to consider the process as taking place in two steps. The most thorough investigation reactions of this type are those involving the formation of Be^8, e.g. $B^{11} + p \rightarrow C^{12} \rightarrow Be^{8*} + \alpha \rightarrow 3\alpha$. Careful analysis of the energy and angular distribution of the α-particles in individual disintegrations indicates the two-step nature of the process [85]. Virtual levels of Ne^{20} (unstable against α-emission) are excited in the reaction $F^{19}(d, n)Ne^{20}$, but the emission of α-particles has not been detected. In general when high bombarding energies are used, especially in the case of deuterons, the emission of a particle from the compound nucleus will leave the residual nucleus in a state unstable against neutron emission $(d, 2n$ reactions). For example, with deuterons of 10 MeV. and a nucleus of medium or large Z, the excitation energy of the compound nucleus will be 20–22 MeV. The most probable energy for neutron emission will be 3–4 MeV. (cf. § 2·2), so that the product nucleus will generally be unstable for neutron and proton emission, although the latter process will be relatively improbable. The interpretation of the experimental observations in terms of energy levels is not so straightforward for those reactions as for the simpler two-body reactions.

CHAPTER IV

RADIATIVE TRANSITIONS

14. INTERPRETATION OF RADIATION PROCESSES

The theoretical interpretation of all processes involving the excited states of nuclei involves the properties of *at least* two nuclear states. Thus in considering the particle groups that can be emitted from a compound nucleus we have to consider both the nature of the state (or superposition of states) of this nucleus itself, and the properties (energy, angular momentum, etc.) of the states of the product nucleus plus the emitted particle which can comprise the final state of the system. A complete interpretation of resonance processes involves, in general, knowledge of three nuclear states: the initial nucleus and incident-free particle, the compound nucleus, and the residual nucleus plus the emitted particle. By means of the dispersion formula it is possible, usually, to condense the relevant properties of these three states into a few empirically determined parameters, viz. the energies of initial and final states of the nuclei and the partial widths for the processes involved. But apart from some general trends (such as the variation of partial width with energy of the dissociated nucleon: $\Gamma \propto \sqrt{W}$, when $\lambda \gg R$), the study of these processes does not provide very direct help in constructing a theory of nuclear structure which will enable these parameters (positions and widths of levels) to be predicted.

The study and interpretation of the emission of electromagnetic radiation between bound nuclear states still involves a knowledge of at least two nuclear states but provides us with somewhat more intimate information about the nature of the nuclear states. In the first place the interaction between nuclei and electromagnetic radiation is of a more familiar type than that between nucleons themselves. It can be assumed to arise from the motion of charged particles in the nucleus in exactly the same way as the interaction between the radiation field and the atom as a whole is used to describe atomic radiative transitions. Just as in the atomic case the interaction between the radiating system and the radiation is weak (i.e. the lifetime for γ-radiation is large compared with the reciprocal of the frequency of the radiation), so that the usual perturbation methods of quantum mechanics that have been applied to the one case can be used for the other. In addition, the wave-length of the emitted radiation is much larger than the radius of the nucleus (cf. atomic case), so that the matrix elements representing the interaction can be expanded in a series with ascending powers of R/λ ($\lambda = 1\cdot3 \times 10^{-10}$ cm. for 1 MeV. quantum energy). An additional feature of great importance in the study of γ-transitions is the strong interaction (in many cases) between the extranuclear electrons and the nucleus in the case of radiative transitions, which is in marked contrast to the minor role played by the extranuclear electrons in the case of nucleon emission. The study of this interaction ('internal conversion') provides very valuable information about the nature of nuclear transitions.

15. MEASUREMENT OF QUANTUM ENERGY

The primary characteristic to be measured for any radiative transition is the quantum energy. Numerous methods have been used for this type of measurement, and new methods are still being developed in attempts to overcome the limitations of existing techniques. The choice of a particular method is determined by the following factors: (i) the magnitude of the quantum energy, (ii) the intensity of the radiation, (iii) the charge of the nucleus from which the radiation occurs, and (iv) the nature of the nuclear process immediately preceding the γ-transition. In view of the importance of γ-energy measurements, we shall discuss briefly the main experimental methods in relation to the above factors.

15·1. Absorption method

The dependence of the absorption coefficient of γ-radiation on energy for all energies arising in nuclear transitions (up to 20 MeV.) is well understood theoretically, and the theory is confirmed by a variety of experimental results [86]. There are two more or less distinct energy ranges to be considered. (a) For energies of about 1 MeV. or more where the absorption is due almost totally to Compton scattering and pair creation the absorption per atom varies monotonically with atomic number. In this energy region, however, the two absorption effects vary in opposite directions with energy, and it is usually necessary to make absorption measurements with *two* or more absorbing materials, since a particular absorption coefficient in a single material may correspond to two energies in the range. The absorption method so employed is capable of fairly accurate results (few per cent) for a homogeneous radiation but is quite inadequate in the case of complex spectra. It is usually necessary to take into account the particular geometrical arrangement employed by calibration with γ-radiation of known energy. (*b*) For energies up to about 0·15 MeV. the absorption coefficient is a discontinuous function of the atomic number of the absorber, increasing slowly with Z up to that value of Z for which the binding energy of the K-electron is greater than the quantum energy. By using a series of absorbers with consecutive values of Z (method of 'selective absorption') the energy of monochromatic soft γ-radiation can be fixed between limits having the ratio of approximately $Z/(Z+2)$. Both types of absorption method, and particularly the latter, since for it only thin absorbers are required, can be used where the intensity is too small for more precise methods to be employed.

15·2. Compton recoil electrons

The energies of the Compton recoil electrons from γ-radiation of energy W_γ form a continuous spectrum with a sharp upper energy limit W_c given by

$$W_c = W_\gamma[2W_\gamma/(m_0 c^2 + 2W_\gamma)].$$

The energy of these electrons, ejected from a thin sheet of material, has been studied by measuring the curvature of their cloud-chamber tracks in a magnetic field (72), a laborious and not very reliable method; by measuring the range of the electrons in aluminium which gives the energy of the most energetic component to an accuracy of 2 or 3 %, but only an indication of components of lower energy (87); by investigation of the electron spectrum in a magnetic focusing spectrograph (94). This last method is capable of high accuracy when sources of high specific activity can be obtained and used with thin sheets of material from which electrons are ejected. In practice a compromise must be struck between resolution and intensity, and in this respect the magnetic lens spectrograph seems preferable to the 180° focusing type where intensity limitations are serious. Spectrograph analysis of Compton electrons can be used for γ-radiation from any nucleus, but for heavy nuclei alternative methods of higher precision are available (see below).

15·3. Photoelectrons

For energies less than 0·5 MeV. the number of photoelectrons ejected from heavy material (usually lead) is larger than the number of Compton electrons; moreover, the photoelectrons form groups with discrete energies equal to $W_\gamma - E_K$, $W_\gamma - E_L$, etc., where E_K, E_L, etc., are the binding energies of K, L, etc., electrons in the atom. This method is useful for investigation of γ-radiation of low energy from light nuclei (where there is not appreciable internal conversion). It is capable of an accuracy of about 1 % with the usual intensities available and can be used to resolve many-lined spectra (38).

15·4. Nuclear photoelectric effect

For quantum energies greater than 2·2 MeV. the photo-disintegration of the deuteron provides a method of measurement whose possibilities have not yet been fully developed. The quantum energy can be deduced from the measurement of proton (or neutron) energy. There is a small uncertainty in the measurement (amounting to 2 % for a quantum energy of 6 MeV.) if the angular relation between the γ-radiation and the emitted particles is unknown. The small cross-section for the nuclear photo-disintegration, of the order 10^{-27} cm.^{2}, provides the main limitation of the technique. Measurement of the neutron energy in the photo-disintegration of Be^9 has also been used to measure γ-energy (89).

15·5. Electron pair creation

Production of electron-positron pairs in a sheet of heavy material (Pb) has been frequently used for measurement of high-energy γ-radiation. The cross-section for pair creation is proportional to Z^2 and increases roughly linearly with quantum energy between 3 and 20 MeV. Measurement of the energies of both the positive and negative electrons of a single pair gives a direct estimate of the quantum energy $(W_\gamma = W_+ + W_- + 2m_0 c^2)$, so that the method is capable of higher accuracy and resolution than the study of Compton recoil electrons, particularly when the electrons are investigated in the cloud chamber and only a limited number of tracks can be measured. Recently a magnetic spectrograph with simultaneous counter-detection of positive and negative electrons has been employed to reduce the labour involved in obtaining statistically reliable results (24).

15·6. Internal conversion

In all the previously mentioned methods, the process by which the emitted quanta are studied occurs at a distance from the radiating nucleus which is large compared with the wave-length of the radiation. Consequently the applicability of these methods does not depend directly on the nature of the nucleus from which the radiation occurs nor on the nature of the transition (electric or magnetic dipole, quadrupole, etc.). If, however, we consider the interaction between an excited nucleus and the inner orbital electrons, i.e. K and L shell electrons, this independence no longer obtains. Therefore a study of the 'internal conversion' electrons, the emission of which depends on the electron wave functions, and therefore on the electric charge of the radiating nucleus, provides information about the multipole nature of the transition as well as its energy.

Corresponding to a particular transition, of energy W_γ, the internal conversion electrons fall into groups having discrete energies $W_\gamma - E_K$, $W_\gamma - E_L$, etc., so that identification of a set of lines of the correct spacing ($E_K - E_L$, etc.) in the spectrum of the conversion electrons enables the corresponding γ-energy to be ascertained. If the process preceding a γ-transition produces N nuclei excited in a particular state, and any one transition from this state can occur either by emission of a γ-quantum or by ejection of an extranuclear electron, then denoting by $\lambda(\gamma)$, $\lambda(K)$, etc., the probabilities (or widths) for the possible processes, the number of K-conversion electrons will be

$$N\lambda(K)/[\lambda(\gamma) + \lambda(K) + \lambda(L) + ...] = N\lambda(K)/\lambda$$

(λ denotes the total probability, per unit time, of the transition occurring). The ratios $\lambda(K)/\lambda = \alpha_K$, $\lambda(L)/\lambda = \alpha_L$, etc., are termed the internal conversion coefficients (i.c.c.) for K, L, etc., electron emission. $\alpha = \alpha_K + \alpha_L + \ldots$ is termed the total i.c.c.

The success of this (the 'natural' line spectrum) method of measuring γ-energies depends primarily on the magnitude of α; α decreases with increasing transition energy (usually as W_γ^{-2} or faster) and is larger for heavy nuclei than light ones. If W_γ is less than $m_0 c^2 (= 0.5 \text{ MeV.})$, then α increases with the multipole order of the transitions, but for larger energies it is only slightly dependent on this characteristic (vide infra). Of course the minimum value of α_K (α_K is usually much larger than α_L, etc., provided $W_\gamma \gg E_K$), which permits successful measurements to be made, depends on such experimental factors as the specific activity of the source of radiation and the effective aperture of the spectrometer, but as a general indication the method is less useful than the previously mentioned spectrographic methods (Compton, photoelectric and pair production) for energies much larger than 1.0 MeV. and for light elements ($Z < 30$). (Also γ-transitions of much lower energy do not frequently occur in light elements.) On the other hand, it is capable of general applicability and high accuracy for most γ-transitions in heavy elements ($Z > 40$). Its high intrinsic accuracy arises from the fact that no *secondary* radiator is required to convert the γ-radiation into measurable electrons; hence the usual compromise between intensity and resolution is, at least in this respect, avoided. The number of internal-conversion electrons is a fixed fraction of the number of γ-transitions; their spread in energy depends only on the thickness of the *source* of γ-radiation.

Most natural radioactive elements can be prepared with high specific activity, and the measurement of conversion electrons in a $180°$ focusing spectrometer using photographic recording is capable of very high accuracy (1 part in 10^3). This technique was developed in the early days of nuclear physics by Ellis, Meitner and others, and still finds useful applications and its accuracy has not been surpassed(90). Under less favourable conditions Deutsch and others have obtained results of somewhat lower accuracy (1 part in 200), using a short magnetic lens spectrograph and counter-detectors. This method has inherent advantages over the semi-circular focusing method, since it permits somewhat larger apertures to be used with the same resolution and also the use of sources of larger superficial area, thereby reducing the absorption and scattering of electrons in the source. A typical spectrum (91) obtained with a magnetic lens spectrometer is shown in fig. 18.

As well as occurring with ejection of a bound orbital electron, a nuclear transition can, if the transition energy exceeds 1 MeV., occur by the emission of an electron pair (i.e. the ejection of an electron from a negative energy state). The probability of this process is too small for it to be utilized as a general method of

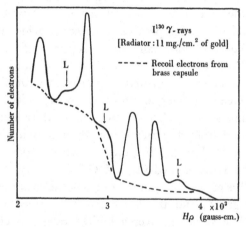

Fig. 18. Photoelectric spectrum of I^{130} γ-rays.
(Roberts, *et al.*, *Phys. Rev.* **64**, 270, 1943.)

energy measurement, although, since the process is practically independent of Z, in marked contrast to K, L, etc., conversion, it may be useful in studying the properties of transitions in light nuclei.

15·7. Crystal spectrometry

The inherent limitations in accuracy and resolution of all methods of γ-spectrometry based on electron energy measurements arise from the loss of energy which occurs when charged particles pass through matter. This limitation is obviated in crystal spectrometry where the wave-length (rather than the quantum energy) of the radiation is measured directly. For energies exceeding 1 MeV. the diffraction angles are too small for the method to be very practical. For smaller energies, crystal diffraction at a plane surface has been used to study nuclear transitions by several investigators (90), the range of application being primarily limited by intensity considerations. Recent developments of focusing crystal spectrometers using curved crystals indicate that the obstacles in the way of wider applicability of the method are not insuperable (91).

15·8. Measurement of γ-intensity

When γ-radiation occurs in competition with other processes, as is usually the case, comparison of the relative probabilities of radiation and of the competitive processes provides information about the radiation widths and other properties of the transition. This usually necessitates measurement of absolute γ-intensity. Most methods make use of a detecting instrument of efficiency less than unity (Geiger counter, ionization chamber) which is calibrated by use of some source of known strength, or whose sensitivity has been estimated from its structure, or by a combination of both methods (72, 39, 93). For example, the number of γ-quanta of energy 6·0 MeV. from the excited state of O^{16} formed in the reaction $F^{19}(p, \alpha) O^{16}$ is equal to the number of α-particles in the associated low-energy group, and the latter can be counted with 100 % efficiency. The variation, over a limited range, of efficiency of detection with quantum energy can then be estimated from the known variation of atomic cross-section for Compton scattering and electron pair production. Radioactive sources of known strength provide means of calibration at low energies. One variation of this method utilizes the fact that positrons, which can be counted readily, when stopped in matter produce two quanta of energy 0·5 MeV., so that the problem of measuring γ-efficiency in this energy region is reduced to one of counting the number of positrons from a source.

If two γ-transitions of (approximately) equal energy occur in rapid succession (10^{-6} sec. or less) the recorded number of coincidences in two γ-ray counters, having overall efficiencies (including solid angle effects) at this energy of η_1, η_2 respectively, is $2\eta_1\eta_2 N$, and the number of quanta recorded by each counter separately $2\eta_1 N$ and $2\eta_2 N$ (N is the number of double transitions per sec.). From measurements of these three quantities η_1 and η_2 can be deduced. If the two γ-transitions are of different energies and two identical counters are used, then the number of γ-γ coincidences will be $2\eta_a\eta_b N$ and the number of γ-quanta recorded by each counter $(\eta_a+\eta_b)N$, where η_a, η_b are the efficiencies of detection for the transitions 'a' and 'b' and N the number of transitions of each type. From an analysis of the absorption curves taken with each counter separately $\eta_a N$ and $\eta_b N$ can be estimated, which, together with the measurement of coincidences, enables η_a, η_b and N to be deduced. Mention has already been made of the use of coincidence counting in ascertaining the correct sequence of transitions in a complex γ-spectrum. In this connexion accurate knowledge of γ-intensities is clearly very useful.

The measurement of the relative probabilities of a γ-transition and of internal conversion is usually made by comparing the number of conversion electrons, in a magnetic spectrometer, with the number of electrons ejected from a thin foil of heavy material (usually Pb) and using combined theoretical and empirical expressions for the probability of electron ejection by a γ-quantum of a particular energy. In this way the need for estimating the aperture of the spectrometer is eliminated. If the γ-transition (or the associated internal conversion) always follows a particular β-transition, then a count of the β-particles provides a measure of the combined number of γ-quanta and conversion electrons, and measurement of the latter provides a measure of the γ-intensity and the I.C.C. With sources of radiation too weak to permit the use of a spectrometer, a modification of this method has been used by Helmholtz in which the ionization produced by conversion electrons has been compared with that produced by photo-electrons ejected by the γ-radiation from material of approximately the same Z as that of the radiating nucleus. The electrons in the two measurements then have similar ionizing properties.

All methods using ejected electrons require a knowledge of the angular distribution of these relative to the γ-radiation. The results are rarely more accurate than about 20 %, but there is seldom any significance (at present) to a more precise comparison with the theoretical estimates, either of internal conversion coefficients or of absolute γ-radiation widths, for nuclear transitions.

16. MULTIPOLARITY OF TRANSITIONS

According to the quantum-mechanical theory of the radiation from a system of charged particles, the probability per second of emission of a quantum of energy in a direction $\varkappa$ and in solid angle $\delta\Omega$, in the transition from state 'a' to state 'b' is (96)

$$P\delta\Omega = \frac{2\pi\nu}{c^3 h} \left| \int \psi_b^* \left[\sum_j e_j/m_j p_j \exp(2\pi i \varkappa . \mathbf{x}_j) \right] \psi_a d\tau \right|^2 \delta\Omega. \quad (16\cdot1)$$

ψ_a, ψ_b are the wave functions for states a, b; e_j, m_j, p_j, $\mathbf{x}_j$ are the charge, mass, component of momentum in the direction of polarization and position of the jth particle, $\varkappa$ is a vector of length ν/c, and in the direction of emission of the quantum. The summation is over all charged particles. For the usual radiation from nuclei $(2\pi i \varkappa . \mathbf{x}_j)$ is very small compared with unity (wave-length $\gg$ nuclear radius), so that the exponential in (16·1) can be expanded in ascending powers of $\varkappa . \mathbf{x}_j$. For the first term in the expansion we have

$$\mathscr{D}\delta\Omega = (2\pi\nu/c^3 h) \left| \int \psi_b^* \left[\sum_j (e_j/m_j) p_j \right] \psi_a d\tau \right|^2 \delta\Omega$$

$$= (2\pi\nu/c^3 h) \left| \sum_j e_j v_j \right|^2_{ab} \delta\Omega, \quad (16\cdot2)$$

where v_j is the velocity component of the jth particle and $||_{ab}$ denotes the matrix element for the transition $a - b$.

Since all matrix elements for the transition $a - b$ contain the time factor $\exp 2\pi i \nu_{ab}$, (16·2) can be written

$$\mathscr{D}\delta\Omega = (2\pi\nu)^3/(c^3 h) \left| \Sigma e_j x_j \right|^2_{ab} \delta\Omega, \quad (16\cdot3)$$

where x_j is the component of $\mathbf{x}_j$ in the direction of polarization.

Integrating this over all directions of polarization and radiation ($\varkappa$) we get for the total probability per unit time D, and hence the radiation width $\Gamma_{\mathscr{D}} (= \hbar\mathscr{D})$

$$\Gamma_{\mathscr{D}} = \tfrac{4}{3}(2\pi\nu/c)^3 \left| \sum_j e_j x_j \right|^2_{ab}. \quad (16\cdot4)$$

This expression, representing only the first term in the expansion of (16·1), gives the radiation width for electric-dipole transitions.

The second term in the expansion of (16·1) gives the probability for quadripole radiation

$$Q\delta\Omega = h^{-1}(2\pi\nu/c)^3 \left| \Sigma e_j x_j (2\pi\varkappa . \mathbf{x}_j) \right|^2_{ab} \delta\Omega. \quad (16\cdot5)$$

This expression is, apart from other factors, smaller than (16·2) by the factor $f = (2\pi\bar{x}/\lambda_{ab})^2$, where $\bar{x}$ is an average distance not larger than nuclear dimensions and λ_{ab} is the wave-length the radiation (e.g. for $\bar{x} = 5 \times 10^{-13}$ cm. and $h\nu = 1$ MeV., $f = 10^{-3}$). Higher terms in the expansion of (16·1) correspond to octupole, etc., transitions, and are smaller than the first (dipole) term in the ratio f^2, f^3, etc. In assessing which of the terms in the expansion corresponds to an actual transition we must consider the different ways in which the matrix elements of the operators $\mathbf{x}, \mathbf{x}(\mathbf{x}.\mathbf{x})$, etc., depend on the properties of the wave functions ψ_a, ψ_b. The most important of these properties (cf. β-transition selection rules) are the symmetry properties defined by the total angular momentum and parity, since these can result in some terms of the expansion vanishing completely. If the two states have total angular momenta $\mathbf{J}_a$ and $\mathbf{J}_b$ and denoting by Δ the quantity $|J_a - J_b|$*, then the first dipole term will vanish unless $\Delta = 0, 1$ and two states differ in parity. (This corresponds to the selection rule for the change in *orbital* momentum in allowed atomic transitions.)

The expression (16·1) represents only transitions due to electric multipoles. Magnetic multipole radiation may also occur, but the probability of such will be less than that of the corresponding electric multipole by a factor $(\bar{v}/c)^2$, where $\bar{v}$ is the typical velocity of the nucleon in the nucleus. The de Broglie wave-length λ_N of a nucleon will be the same order of magnitude as the nuclear radius R, and therefore denoting by ϵ the energy of a nucleon we have

$$\bar{v}/c \sim \epsilon/m\bar{v}c \sim (\epsilon/hc)\,\lambda_N \sim R/\lambda,$$

so that magnetic dipole radiation will generally be of comparable importance with electric quadrupole and so on. The general rules for identifying the first non-vanishing, and therefore the most important terms in the expansion of (16·1), and the corresponding expression for magnetic radiation are given in table 2 (97). For example, for $\Delta = 2$ and parity change, the first non-vanishing terms are electric 2^3 (octupole) and magnetic 2^2 (quadrupole), and these two terms will, in general, be of comparable importance. But for $\Delta = 2$ and no parity change, of the first two non-vanishing terms electric quadrupole will be by far the more important. Generally for

* J denotes the integer related to $\mathbf{J}$ by $|\mathbf{J}| = \hbar\sqrt{\{J(J+1)\}}$.

transitions of (even Δ, $\neq$) and (odd Δ, $=$) type magnetic and electric radiation will be of comparable importance; and for (even Δ, $=$) and (odd Δ, $\neq$) electric transitions will be more probable by a factor of order f^2.

Angular momentum and parity are not the only properties of the nuclear states which determine the magnitude of the matrix elements. For example, if the centroids of charge and mass coincide for all states (which may be approximately true in a nucleus consisting of a uniform distribution of protons and neutrons), then the dipole term in (16·1) vanishes whatever the value of angular momentum and parity change (98). In fact the probabilities of

TABLE 2*

Parity difference	Δ even	Δ odd
No change ($=$)	2^Δ electric $2^{\Delta+1}$ magnetic	$2^{\Delta+1}$ electric 2^Δ magnetic
Change ($\neq$)	$2^{\Delta+1}$ electric 2^Δ magnetic	2^Δ electric $2^{\Delta+1}$ magnetic

electric-dipole and quadrupole radiation from a nucleus are of comparable magnitudes (vide infra) despite the expected ratio of f, a result which may be interpreted in the above manner or on the basis of approximate symmetry properties proposed by Wigner (§ 19·1).

16·1. Multipolarity and internal conversion

It is not possible to calculate the absolute probability of radiative transitions from (16·1), etc., without some knowledge of the matrix elements, and estimates of these require some nuclear model. However, the radiation field due to any postulated strength of multipole oscillator can be written down, and the interaction of such a field with the atomic electrons can be calculated (in principle)

* In this table it is assumed that both J_a and J_b are sufficiently large for the angular momentum required for the particular multipole radiation to be available by suitable (vectorial) combination of $\mathbf{J}_a$ and $\mathbf{J}_b$, i.e. it is assumed that the series expansion of the matrix elements extends as far as the terms listed. This is not the case if either J_a or J_b is zero. For example, if $J_a = 0$, $J_b = 1$ and the states have the same parity, electric quadrupole radiation cannot occur, since this would carry away two units of angular momentum which is not available. Only magnetic dipole radiation can occur. If $J_a = J_b = 0$ all terms in the expansion vanish.

from the known wave functions of the latter. Hence the I.C.C. can be calculated independently of the *magnitude* of the nuclear matrix element but with some assumption about the multipole nature of the transition. Explicit expressions for the I.C.C. applicable to all values of Z, transition energies and multipole types have not been obtained on account of the mathematical complexities, but calculations have been made with varying degrees of approximation in

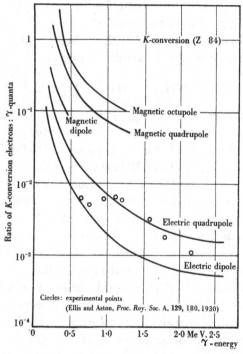

Fig. 19. Internal conversion coefficients ($Z=84$).

particular cases. The most precise calculations have been made for the natural radioactive elements, the only approximations being the neglect of the screening of the nuclear charge by the extra-nuclear electrons and the assumption that the radiation field retains its characteristic form (dipole, etc.) even for distances from the centre of the atom less than the nuclear radius (i.e. radiation from an ideal point multipole). The theoretical values (99) for α_K in the case $Z = 84$ are shown in fig. 19 (α_K is roughly proportional to Z^3). Conversion in the L shell is approximately one-seventh of

K-conversion for transition energies much larger than the K-binding energy, and this ratio is about the same for dipole and quadrupole radiation. For transitions of energy less than the K-binding energy α_L approaches unity; for low energies generally the ratio α_L/α_K depends on the multipole nature. Calculations of internal conversion in light elements have been made only in some limiting cases. Hebb and Uhlenbeck [100] have obtained an explicit expression for $\alpha_K/(1-\alpha)$ by using non-relativistic electron wave functions, which is reasonable for transition energies small compared with 0·5 MeV. and $Z < 40$. Their result (applicable to electric transitions only) is

$$\alpha_K/(1-\alpha) = 32\pi(\tfrac{1}{137})^{2\Delta+1} \prod_{l=1}^{\Delta}(2l-1)\frac{1}{W}\left(\frac{Z}{2W}\right)^{2\Delta}K(\Delta), \quad (16\cdot6)$$

where W, the transition energy, is in units of $m_0 c^2$ and $K(\Delta)$ is a complicated function of Z, the electron momentum and Δ, but is not the dominant factor in expression (16·6). This result illustrates the suggestion of Weizsäcker [101] that for low-energy transitions of high multipolarity, electron conversion is far more probable than γ-emission. The ratio of the probabilities of K-electron and γ-emission contains a factor $(\lambda/a_K)^2$, where a_K is the Bohr radius of the orbit of the K-electron, so that if $\lambda \gg a_K$, the ratio of K-conversion to γ-radiation will increase rapidly with multipole order. ($\lambda \gg a_K$ is equivalent to the transition energy W, in MeV., satisfying $W \ll Z/40$.) The calculations have been extended by Dancoff and Morrison [102] and Hebb and Nelson [103] to include magnetic-multipole transitions and L-conversion. For $W \ll m_0 c^{2*}$ explicit formulae are obtained for $\alpha_K/(1-\alpha)$ (i.e. the ratio of the number of K-electrons to γ-quanta), if the binding energy of the electrons is negligible in comparison with W (i.e. $Ze^2/Wa_K \ll 1$). With this approximation the results are

Electric transition:

$$\alpha_K/(1-\alpha) = (2Z^3/W^3)(\tfrac{1}{137})^4\{(W+2)/W\}^{\Delta-\frac{1}{2}}\left\{\frac{(\Delta+1)\,W^2+4\Delta}{\Delta+1}\right\},$$

Magnetic transition:

$$\alpha_K/(1-\alpha) = (2Z^3/W^3)(\tfrac{1}{137})^4\{(W+2)/W\}^{\Delta+\frac{1}{2}}.$$

$$(16\cdot7)$$

---

* Relativistic treatment is essential for magnetic multipole conversion, since this is essentially an electron spin phenomena.

W is again in units of $m_0 c^2$ and Δ represents the multipolarity of the transition.

Using non-relativistic wave functions and assuming the wavelength of the γ-radiation to be much *larger* than *atomic* dimensions (i.e. the radius of K, L orbits) Hebb and Nelson obtained the results shown in fig. 20 for the ratio α_K/α_L in electric transitions. The values are correct for $Z = 35$, but on account of electron screening of the nuclear charge the values are too low for $Z < 35$ and conversely. For $25 < Z < 50$ they are correct to within about 20 %.

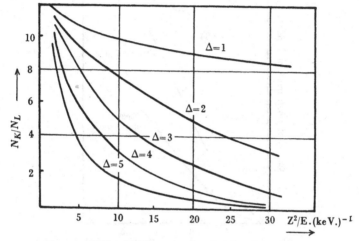

Fig. 20. Ratio of K/L conversion for electric multipoles. (Hebb and Nelson, *Phys. Rev.* **58**, 486, 1940.)

For magnetic radiation plane-wave functions were used for the ejected electrons, i.e. the binding energy of the electrons assumed to be much less than transition energy, and with this approximation (equivalent to $Z^2/W \ll 70$), the ratio α_L/α_K is given by

$$(\alpha_L/\alpha_K)_{\text{Mag.}} = \frac{1}{8}\left\{1 + \left(\frac{Z}{137}\right)^2 \left(\frac{W+2}{4W}\right)\right.$$
$$\left. \times\left[\frac{\Delta+1}{2\Delta+1} + \frac{\Delta(2\Delta+1)}{4}\left(\frac{2\Delta-1}{2\Delta+1} - \frac{W}{W+2}\right)^2\right]\right\}. \quad (16\cdot8)$$

The relative probabilities of pair production in the field of the radiating nucleus and of radiation itself $\alpha_\pi/(1-\alpha)$ has been calculated by Jaeger and Hulme (104) and their results are shown in fig. 21.

In marked contrast with K, L, etc., conversion, the value of α_π is practically independent of Z, is greatest for electric dipole radiation and *increases* with increasing energy. One special case of interest is the transition between two states with $J = 0$. In this case no γ-radiation can occur, but if the states have the same parity the transition can take place by internal K-conversion or pair creation. If the parities of the two states are different no single process of an

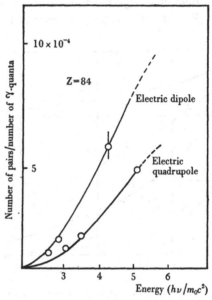

Fig. 21. Internal conversion coefficient for pair production. Circles: experimental (Latyshev, *Rev. Mod. Phys.* **19**, 132, 1947). Curves: theoretical (Jaeger and Hulme, *Proc. Roy. Soc.* **148**, 708, 1935).

electromagnetic nature can occur, but two-quantum emission is a (theoretically) possible process of long lifetime [105].

17. COMPARISON WITH EXPERIMENT

17·1. Dipole transitions

If we consider a transition of low multipolarity and high energy, so that the internal conversion coefficients are all small, then the total width for such a transition will be effectively the radiation width. An estimate of this quantity can be made, for the electric-dipole case, by substituting in (16·4) for $|\Sigma e_j x_j|^2$ the matrix element of a single electron charge moving in an ordinary three-dimensional oscillator potential of appropriate size (for a frequency ν), namely, $(3\hbar^2 e/2m\nu)$. For a γ-transition of $1\cdot0$ MeV. this would give a width of the order 10 eV. (half-life for γ-radiation: $0\cdot4\times10^{-16}$ sec.). A similar result is arrived at by replacing $|\Sigma e_j x_j|$ by the product of a single electronic charge and the nuclear radius, (eR).

The radiation width for *heavy nuclei* has been measured, in the case of the natural radioactive α-emitting nuclei Ra C$'$, Th C$'$, by comparing the relative numbers of 'long-range' α-particles and γ-quanta emitted from some excited levels (cf. §4·2). The absolute probability of α-particle emission can be derived from a semi-empirical extrapolation of the relation between lifetime and α-particle energy (Geiger-Nuttall relation). Widths of the order of 10^{-3} eV. (lifetime 10^{-12} sec.) for γ-transitions of the order 1 MeV. energy are obtained in this way, a result which is about 10^{-4} times less than indicated by the simple hypothesis above. Since the dipole radiation width is so small, it is possible that dipole and quadrupole radiation will be of comparable importance if, as is probable, the factors reducing the dipole matrix elements to a value small in comparison with (eR) do not affect the order of magnitude of the quadrupole matrix elements. This is borne out by the experimental observation that for γ-transitions in natural radio-active elements, dipole and quadrupole transitions, as identified by measurements of the I.C.C. (cf. fig. 19), appear about equally often.

For *light nuclei* the best information about γ-radiation widths comes from proton capture measurements (§9·1).* If, for example, we assume the γ-transition of 17 MeV. in the process Li$^7(p,\gamma)$Be^8 is dipole, the value obtained for $|\Sigma e_j x_j|$ is $e\times0\cdot6\times10^{-14}$ cm., which is rather less than might be expected from nuclear dimensions. Replacing $|\Sigma e_j x_j|$ by the product of a single electronic charge and the nuclear radius, 3×10^{-13} cm., gives a value for the width of about 300 eV. (i.e. 30 times too large). Measurement of K-internal-

---

* Attempts have recently been made to measure the half-life (and hence the radiation width) of γ-transitions from *bound* states by examining the Doppler broadening of the radiation when the radiating nucleus is produced as a recoiling residual nucleus in a nuclear reaction. This method has shown, for example, that the γ-transition from the excited state of Li^7 at $0\cdot48$ MeV. has a half-life of less than 10^{-12} sec. (Elliott and Bell, *Phys. Rev.* **74**, 1869, (1948).)

conversion cannot, unfortunately, be used in the case of such light elements to identify the multipolarity of the transitions.

Measurements of the I.C.C. for pair production is a feasible method of investigation in these cases. So far the only measurement of this quantity in light elements shows marked disagreement with theory (106) (probably attributable to experimental inaccuracies),* although measurements in heavy nuclei are in agreement with theory.

The γ-radiation widths of the resonance levels observed in slow-neutron capture (§ 8) mostly lie between 0·05 and 0·2 eV. and do not appear to vary much between $Z \sim 48$ and $Z \sim 80$. These widths are compounded of the partial radiation widths for transitions to all the combining levels of lower energy than the resonance level, and these will be very numerous for such heavy nuclei. Since, according to (16·3), the partial width for a transition contains the factor ν^3, the lower levels give an important contribution to the total radiation width despite the higher density of levels near the resonance level. Denoting the individual levels of excitation energy E_k by the suffix k, and the excitation energy of the resonance level by E_r (6–8 MeV. for the relevant nuclei), then the total radiation width, assuming electric-dipole transitions, is given by

$$\Gamma_\gamma (\text{total}) = \sum_k \Gamma_{rk} = \sum_{k=0}^{r} \frac{4}{3c^3} [(E_r - E_k)/\hbar]^3 \, |X|_{rk}^2, \quad (17\cdot1)$$

where $|X|_{rk}$ is the matrix element $|\sum_j e_j x_j|_{rk}$ for the transition $r-k$, and the summation is over all levels between the resonance level 'r' and the ground state 'o'. To obtain an order of magnitude for Γ_γ (total), we can replace the Σ by $\int$, assuming a continuous density $\rho(E)$ of levels of excitation energy E, and replacing $|X|_{rk}$ by an average value.

$$\Gamma_\gamma (\text{total}) = \frac{4}{3c^3} |\bar{X}|^2 \left(\frac{E_r}{\hbar}\right)^3 \int_0^{E_r} [(E_r - E)/E_r]^3 \rho(E) \, dE. \quad (17\cdot2)$$

The integral in (17·2) represents the ratio of the total radiation width of the resonance level to the partial width for a single γ-transition to the ground state. Its value depends only on the

* More recent measurements do not disagree with theoretical estimates. (Dougherty, Hornyak, Lauritsen and Rasmussen, *Phys. Rev.* **74**, 1727 (1948).)

assumptions made about $\rho(E)$, but with any reasonable assumption about the spacing of energy levels it is likely to be 100 or more for $Z > 50$. This, together with the experimentally observed widths, implies a partial γ-radiation width for the single transition $r - 0$ of the order 10^{-3} eV. or a value for $|\overline{X}|$ of $e \times 10^{-15}$ e.s.u.cm. This may be a low estimate, since the angular momentum of many of the lower states may be too large to permit electric-dipole radiation, but the result confirms the general indication that the dipole matrix elements are always very much smaller than the simple product of nuclear radius and electron charge, and that, consequently, quadrupole radiation may be of comparable probability.

17·2. Higher multipoles

A system with a uniform but non-spherical mass and charge distribution will have zero electric-dipole moment, but the electric quadrupole and higher moments will not be zero. It appears reasonable to take as a model of the nucleus such a system of uniformly distributed charges (i.e. with the centroids of neutrons and protons coincident) and consider a transition of low energy as corresponding to a change in shape of the nucleus in which the centroids of charge and mass remain stationary and coincident. With such a model (cf. §20·5) the ratio of quadrupole to octupole and higher matrix elements will be given correctly by the factor (inter alia) of $(2\pi R/\lambda)^2$. Despite, then, the incompleteness of the theoretical interpretation of the small *dipole* matrix elements, we can make estimates of the radiation widths for higher multiple transitions with some confidence, which is sustained by comparison with experimental data.

For quadrupole radiation, the expression for the width corresponding to (16·4) is

$$\Gamma_Q = \tfrac{1}{4}(2\pi\nu/c)^5 |Q|^2_{ab},$$

where $|Q|$ is the value, averaged over all directions of radiation, of the quadrupole term in (16·1), i.e. $|\sum_j e_j x_j(\mathbf{n}.\mathbf{x}_j)|^2$ ($\mathbf{n}$ is a unit vector in the direction of propagation of the radiation). For a transition energy of 1 MeV. a width of 10^{-3} eV. indicates a quadrupole moment of $e \times 0.3 \times 10^{-24}$ e.s.u.cm.^{2} which is of the same order of magnitude as eR^2.

Several attempts have been made to obtain the radiation widths of multipoles higher than electric dipole, by considering in a semi-classical way the oscillations of a specific nuclear model.

Hebb and Uhlenbeck (100) have based estimates on an α-particle model; Bethe and others on the liquid-drop model, in which transitions are represented by surface oscillations of an incompressible liquid. All the calculations give similar results for the dependence of the radiation width on energy and multipolarity; namely, the width decreases with increasing multipole order, Δ, and with increasing wave-length of the radiation due primarily to the factor $(2\pi R/\lambda)^2$.

· For example, Bethe (107) gives the following expression for the radiation width:

$$\Gamma_\gamma = (2\pi/\lambda)^{2\Delta+1} e^2 R^{2\Delta}/(\Delta!)^2, \qquad (17\cdot3)$$

where R is a length with order of magnitude equal to the nuclear radius. With this expression, for a transition of energy 5×10^4 eV. and $\Delta = 4$, $R = 10^{-12}$ cm., Γ is 10^{-21} eV. corresponding to a half-life of 10 days! Many such γ-transitions with measurable half-lifes ($\geqslant 10^{-7}$ sec.) have in fact been observed in medium and heavy nuclei and are usually termed 'isomeric transitions'. It is assumed that they occur between a low-lying (probably the first excited state) level of the nucleus having angular momentum very different from the ground state, and the ground state. No competitive transitions* are possible under these circumstances.

For these transitions the radiation width can be deduced directly from the measured half-life of the transition, but before a comparison can be made with theoretical formulae, it must be remembered that the conditions for long half-life (large Δ and small energy) are just those for very large internal conversion coefficients (cf. §16·1), so that the width deduced from the half-life must be compared not with the radiation width, Γ_γ, but with the total transition width $\Gamma = \Gamma_\gamma[1/(1-\alpha)]$, where α is the total (K, L, etc.) I.C.C. For these low-energy transitions the approximations (neglect of binding energy) made in obtaining the expressions (16·7) may be invalid, but for electric multipole transitions, (16·6), in which non-relativistic wave functions were used, may be a reasonable approximation

---

* Except in some instances when β-decay can occur from both isomeric and ground states, but this is also a slow process.

for medium nuclei ($Z < 50$). For magnetic transitions relativistic wave functions must be used in calculating the I.C.C., but the results of Dancoff and Morrison in which the electron binding energy is neglected may be in considerable error for low-energy transitions.

Despite these approximations, measurement of the half-life of these γ-transitions provide a good estimate of Δ once the transition energy has been measured, since a change in Δ of unity produces in a typical case a change in half-life of a factor of the order 10^4. The

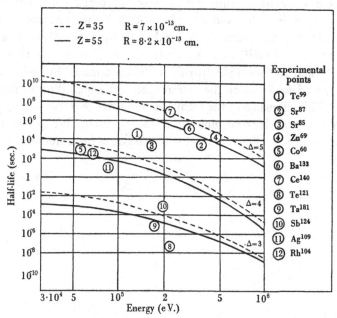

Fig. 22. Life-time of isomeric transitions.

value of Δ can also be estimated from two other measurements: the I.C.C. for K-electrons and the ratio of K/L conversion (although a unique value of the I.C.C. is not provided theoretically by a knowledge of Δ and the transition energy, since the transition may be mixed electric and magnetic and the I.C.C. will depend on the ratio of the two components). Several transitions have been studied in this way by Helmholtz [93] and others, and the results indicate that in many cases the radiation is neither purely electric nor purely magnetic, and that the theoretical estimates, both of I.C.C. and radiation width, are scarcely refined enough for an accurate comparison with experiment. For example, Zn^{69} shows an isomeric level of energy $0\cdot439$ MeV. with half-life $13\cdot8$ hr. Theoretical estimates of the half-life gives a value about 10 times too large for $\Delta = 5$, and 10^4 too small for $\Delta = 4$. The approximately measured i.c.c. indicate either 2^5-pole electric or 2^4-pole magnetic radiation.

Fig. 22 shows the variation of half-life with energy for transitions in nuclei with $Z = 35$ and with $Z = 55$. The transitions are assumed to be electric multipole, and the expression ($17\cdot3$) is used for the γ-radiation width and expression ($16\cdot7$) for internal conversion. Some observed energies and half-lifes are shown for comparison (108). γ-transitions with half-lives of approximately 5×10^{-8} sec. (Te^{121}) (109), 2×10^{-5} sec. (Ta^{181}) (110), 7×10^{-5} sec. (Ce^{141}) and 10^{-3} sec. (Sb^{124}) (111) have been examined by measuring the time delay between the γ-transition and a preceding β-transition. This extension of the technique indicates that the distinction between isomeric and ordinary γ-transitions is simply a question of convenience in relation to available methods of measurement.

CHAPTER V

NUCLEAR SPECTRA AND THEIR INTERPRETATION

18. SCOPE OF THEORETICAL INTERPRETATION

Any attempt to interpret or even to classify the extensive experimental information about the energy levels of individual nuclei is beset by two major difficulties. In the first place the experimental results, even for those nuclei which have been studied most thoroughly, cannot be relied upon to give a *complete* picture of the spectrum of nuclear levels, either because only a limited region of energy has been accessible to investigation or because the evidence, even in a limited region, is only fragmentary. Moreover, new experimental material is being produced so rapidly that any straightforward catalogue of data would be out of date and possibly unrepresentative by the time it could be assembled. Secondly, there is no comprehensive theoretical framework within which the detailed characteristics of particular nuclei can be fitted, and which would serve as a guide in marshalling the data in the most significant order. We shall therefore present here a brief account of the main theoretical approaches to the problem of nuclear spectra together with some experimental data chosen for comparison with particular theoretical predictions.

The complicated nature of the 'many-body' problem coupled with the still uncertain quantitative particulars of nuclear forces have resulted in limited progress in the theory of nuclear spectra. For very light nuclei H^2, H^3, He^3 and He^4, either exact solutions are possible (H^2) or attempts can be made to find explicit wave functions for the nuclei which can be used, together with some suitably adjusted combination of ordinary and exchange forces, to give agreement with observed values of the energies of the *ground states* of these nuclei [112]. The usual mathematical technique employed is to incorporate some arbitrary parameters in the wave function (and sometimes also in the Hamiltonian expression) and apply the variational principle, i.e. adjust the arbitrary parameters so as to produce a minimum value for the expression for the total energy obtained by using these parameters in the wave function and potentials. In principle it is possible, after having ascertained a suitable form for the potentials, to apply the variational principle to find wave functions, and hence energies for stationary states other than the ground state. This procedure meets with serious difficulties however, since, first, the wave functions must be restricted to those orthogonal to the ground-state wave functions, and secondly, the minimum will no longer be an absolute one but will apply only to a local range of variations. In view of the limited success of the variational methods applied to ground-state calculations, and their general lack of refinement, it is unlikely that a detailed interpretation of spectra, even of light elements, will be developed along these lines.*

In order to develop a more productive theory it is necessary to make some simplifying approximations about the nature of the forces between nucleons or to adopt some simple model of the structure of the nucleus, in which assumptions about nuclear forces may be indirectly incorporated. These two types of simplification lead to theoretical results which are applicable to essentially different ranges of phenomena.

19. SYMMETRICAL FORCE THEORY

Measurement of the scattering of high-energy protons by protons and neutrons by protons seems consistent with the hypothesis that the specifically nuclear forces (i.e. other than Coulomb repulsion) between a neutron and a proton (n, p) and between two protons (p, p) are equal. Direct information about (n, n) interaction is not available, but it is convenient and consistent with indirect information to assume that it is equal to the (n, p) and (p, p) interaction. Furthermore, it is in agreement with observation to assume that in actual nuclei, and particularly light nuclei, that the role of the Coulomb forces is a minor one. With these assumptions, the evaluation of the energies of the nuclear stationary states can be made by considering first the eigen-functions of the Hamiltonian

---

* Calculations of this type have been more useful in providing qualitative results, e.g. indicating the finite range of nuclear forces and the existence of proton-proton and neutron-neutron forces (the former in agreement with direct experiment), than in providing quantitative predictions about energies of bound states.

obtained by assuming only nuclear forces (n,p), (p,p), (n,n), and these all equal (symmetrical Hamiltonian, $\mathscr{H}_\tau$), and then considering the Coulomb energy as a (small) perturbation. If this perturbation energy is small compared with the separation of the eigenvalues of the (unperturbed) symmetrical Hamiltonian, $\mathscr{H}_\tau$, then the Coulomb perturbation will not cause much mixing of the eigen-states of $\mathscr{H}_\tau$. Since $\mathscr{H}_\tau$ is independent of the charge on the nucleus, then in so far as may be compatible with the exclusion principle, eigen-states of $\mathscr{H}_\tau$ may correspond to states of *different*, *isobaric* nuclei, i.e. nuclei having the same number of nucleons but different charges. The actual energies of these corresponding states of the different isobars will, of course, differ by the small Coulomb perturbation.

A further assumption which it is convenient to make is that the forces between nucleons is spin-independent. This is known not to be exactly true (cf. n, p scattering, quadrupole moment of the deuteron), but it seems reasonable to assume that the spin-dependent part of the interaction is much smaller than the main spin-independent. With this further approximation, a Hamiltonian function $\mathscr{H}_{\tau\sigma}$ can be used, in which both spin and charge dependence are ignored, to define a set of eigen-states, and one in which the ignored factors are taken into consideration as perturbations. A particular eigen-value of $\mathscr{H}_{\tau\sigma}$ will then correspond to a whole range of values of nuclear charge and spin (for a fixed number of nucleons), the range of values of these variables (not independent!) being governed by the exclusion principle. The group of 'states' (in a generalized sense since it refers to a set of isobaric nuclei) into which a single degenerate level of the approximate Hamiltonian $\mathscr{H}_{\tau\sigma}$ is split under the influence of the perturbations, has been termed by Wigner a 'super-multiplet' in analogy to the multiplets of atomic spectra. (In this latter case the approximate Hamiltonian ignores spin-orbit interactions which are subsequently introduced as perturbations.) The development of the formal theory of the symmetry properties of nuclear states based on these approximations has been made by Wigner[113], and a comprehensive review of its application to the interpretation of a variety of nuclear properties has been written by Wigner and Feenberg[114].

19·1. Wigner's theory of 'super-multiplet' structure

The state of charge of a nucleon can be represented, in analogy with the ordinary spin variable, by an 'isotopic spin' variable, τ, assuming the values $+1$ (neutron) and -1 (proton), and the wave function of a nuclear system, comprising a set of isobaric nuclei with A nucleons, can be written as $\psi(r_1\sigma_1\tau_1, r_2\sigma_2\tau_2, ..., r_A\sigma_A\tau_A)$, where r, σ, τ refer to space, ordinary spin, and isotopic spin variables. The exclusion principle requires that

$$\psi(...r_j\sigma_j\tau_j...r_k\sigma_k\tau_k...) = -\psi(...r_k\sigma_k\tau_k...r_j\sigma_j\tau_j...).$$

The quantity $\frac{1}{2}\sum_i\tau_i$ ($\equiv T_\zeta$) defines the total charge, but for a particular eigen-value of the approximate Hamiltonian $\mathscr{H}_\tau$ (or $\mathscr{H}_{\sigma\tau}$) characterized by the quantum number T, T_ζ can assume the $2T+1$ values ranging from $-T$ to $+T$. This set of degenerate states (compare the $2S+1$ members of a multiplet in atomic spectra), which refers to *different* isobars, is called an isobaric or 'T' multiplet. In this approximation T, as well as J (the total angular momentum) and the parity, is a good quantum number (and the same for all members of the T-multiplet). It will remain approximately so when the Coulomb perturbation is taken into account provided this perturbation energy is small compared with the separation of the eigen-values of $\mathscr{H}_\tau$.

The energies of both unperturbed and perturbed states of a T-multiplet may be represented as in fig. 23, where the *binding energy* is plotted as a function of T_ζ.

Particular simple cases of T-multiplets are those for $T=0$ and $T=\frac{1}{2}$. The former corresponds only to states of nuclei with equal numbers of neutrons and protons, e.g. the ground states of nuclei of the type He^4, Be^8, C^{12} and for these T-multiplets there is no (approximate) degeneracy with respect to T_ζ. For $T=\frac{1}{2}$, T_ζ can have two values $\pm\frac{1}{2}$, and the ground states of nuclei with one odd proton and neutron (up to mass number 41) are of this type (e.g. Be^7, Li^7, C^{11}, B^{11}, N^{13}). That the ground states of these nuclei do belong to the same T-multiplet is confirmed by the fact that the difference of binding energies of the isobaric pairs amongst them (e.g. Be^7, Li^7) can be accounted for entirely by the Coulomb perturbation energy.* There is, however, no experimental evidence to suggest that the level spectra of these isobaric pairs are similar. This may be due to the fact that the spectra of the $T_\zeta = -\frac{1}{2}$ members of the T-multiplets have been subject of very little investigation; and furthermore, these nuclei are only stable against dissociation of a heavy particle by a few MeV., which is the same order of magnitude as the excitation energy of the first few levels of these light nuclei. Fig. 24b illustrates the known properties of the (C^{13}, N^{13}) isobaric pair for which the experimental knowledge is about as complete as for any of the light nuclei.

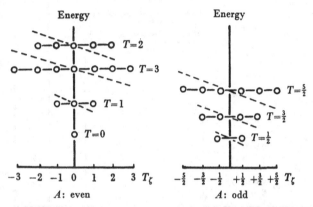

Fig. 23. T-multiplets for nuclei of even and odd mass. (Circles denote some possible nuclear states; dotted lines indicate effect of Coulomb forces.)

For nuclei of even mass number, but not of the $4n$ type, the ground state may belong to a $T = 0$ or $T = 1$ multiplet. In the isobaric group C^{10}, B^{10}, Be^{10}, the ground states of Be^{10} and C^{10} are probably members of a $T = 1$ multiplet ($T_\zeta = \pm 1$), and that of B^{10} belongs to the single state of a $T = 0$ multiplet ($T_\zeta = 0$). If this is so then there should exist a state of B^{10}, belonging to the $T = 1$ multiplet (with $T_\zeta = 0$), having energy equal to the mean of the ground-state energies of Be^{10} and C^{10}. There is some experimental evidence for this (fig. 24a).

The more general symmetry properties of the groups of (degenerate) eigen-states of $\mathcal{H}_{\sigma\tau}$ are very much more involved than those of $\mathcal{H}_\tau$, since the ordinary and isotopic spin degeneracies are

* In these particular cases equality of (n, n) and (p, p) forces alone would yield this result.

not independent but are related through the exclusion principle; and the total spins both of protons and neutrons are separately good quantum numbers for eigen-states of $\mathcal{H}_{\sigma\tau}$. Wigner characterizes the symmetry properties of eigen-states of $\mathcal{H}_{\sigma\tau}$ by means of four 'partition quantum numbers', $\Lambda_1 \leqslant \Lambda_2 \leqslant \Lambda_3 \leqslant \Lambda_4$, which are related by $\sum_{1234} \Lambda = A$, the number of nucleons, so that only three *independent* quantum numbers are necessary for description of the symmetry.

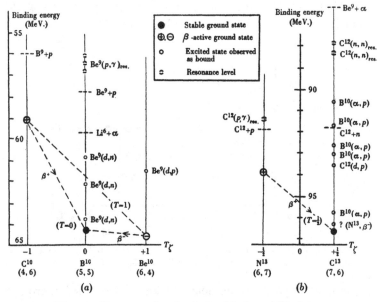

Fig. 24. Known nuclear levels for (a) $A = 10$, (b) $A = 13$.

The significance of the partition quantum numbers can be seen most easily in the simple case of a system of nucleons moving independently in a fixed field of force (e.g. potential well model). If in this case $\lambda_1, \lambda_2, \lambda_3, \lambda_4$ refer to the *actual* number of nucleons with $\sigma_z = 1$, $\tau_\zeta = 1$; $\sigma_z = -1$, $\tau_\zeta = -1$; etc., then $\Lambda_1 \ldots \Lambda_4$ are defined as the highest set of λ's of *any* state belonging to the same eigen-value of $\mathcal{H}_{\sigma\tau}$ as the particular one having the distribution $\lambda_1 \lambda_2 \lambda_3 \lambda_4$. The 'higher' of two sets λ, λ' is the one with the largest value of λ_4; or if $\lambda'_4 = \lambda_4$, with the larger λ_3; or if $\lambda_3 = \lambda'_3$, with the larger λ_2. It is clear that in this notation the lowest states will have all four λ's approximately equal, i.e. all nucleons in the lowest states permitted by the exclusion principle. By choosing as *independent* partition numbers, the quantities

$$P = \tfrac{1}{2}(\Lambda_4 + \Lambda_3 - \Lambda_2 - \Lambda_1), \quad P' = \tfrac{1}{2}(\Lambda_4 - \Lambda_3 + \Lambda_2 - \Lambda_1),$$
$$P'' = \tfrac{1}{2}(\Lambda_4 - \Lambda_3 - \Lambda_2 + \Lambda_1),$$

the symmetry of the state of $\mathscr{H}_{\sigma\tau}$ is designated by quantities which do not involve the number of nucleons, A, explicitly; are the same for two values of A differing by an integral number times 4; and are smallest for the lowest energy levels. These three numbers, P, P', P'', determine the fine-structure pattern of the set of states (super-multiplet) arising from a single highly degenerate level (one eigen-value of $\mathscr{H}_{\sigma\tau}$) which splits up under the influence of charge-dependent (e.g. Coulomb) and spin-dependent perturbation forces P, P', P'' determine the range of the variables

$$S_z(=\tfrac{1}{2}\Sigma\sigma_z), \quad T_\zeta(=\tfrac{1}{2}\Sigma T_\zeta) \text{ and } Y_\zeta(=\tfrac{1}{2}\Sigma\sigma_z\tau_\zeta)$$

(the summation is over all nucleons) occurring in a particular super-multiplet. There is, in general, an additional degeneracy arising from the occurrence of the same set of values of S_z, T_ζ, Y_ζ more than once in a single super-multiplet.

The possible values of the P's depend on which of the groups $4k$, $4k+1$, $4k+2$ to which A belongs. For A even all P's are integral; A odd, half-integral, $\Sigma P = 2f - \tfrac{1}{2}A$ (f is an integer) in all cases and the maximum value of the P's is determined by the relations

$$P \geqslant P' \geqslant P'', \quad P' + P'' > 0 \text{ and } P + P' + P'' \leqslant A.$$

The properties of a super-multiplet are given by Wigner by reducing to sets of values of the variables (S, T) all the possibilities which arise from a given set of partition numbers, P. The values of S_z which occur are the $(2S+1)$ values from S to $-S$ and similarly for T_ζ. The occurrence of the pairs (S, T) always implies also (T, S) and the maximum S or T is equal to P. This is illustrated in table 3, showing the range of S, T for the few lowest super-multiplets.

The eigen-functions of $\mathscr{H}_{\sigma\tau}$ corresponding to a particular super-multiplet will all have a definite value of L, total orbital momentum, and although for a particular nuclear model there may be some correlation between L and the type of super-multiplet, the super-multiplet *classification* is not explicitly dependent on L. Consequently the perturbation introduced by spin-dependent forces can be regarded as removing the degeneracy of the different (S, T) combinations of the same super-multiplet, and the fine structure introduced ($2S + 1$ splitting except when $L < S$) will depend on both L and S, where L is characteristic of the super-multiplet as a whole. Fig. 25 shows the level patterns for three typical super-multiplets $(2, 0, 0)$, $L = 1$; $(1, 1, 0)$, $L = 2$; and $(0, 0, 0)$, $L = 0$.

TABLE 3

Type $4k$		Type $4k+2$	
(P, P', P'')	(S, T) or (T, S)	(P, P', P'')	(S, T) or (T, S)
$(0, 0, 0)$	$(0, 0)$	$(1, 0, 0)$	$(0, 1)$
$(1, 0, 0)$	$(0, 1)(1, 1)$	$(1, 1, \pm 1)$	$(0, 0)(1, 1)$
$(2, 0, 0)$	$(0, 0)(1, 1)(0, 2)$	$(2, 1, 0)$	$\{(0, 1)(1, 1)$
$(2, 1, \pm 1)$	$(0, 1)(1, 1)(1, 2)$		$\{(0, 2)(1, 2)$

Type $4k \pm 1$	
(P, P', P'')	(S, T) or (T, S)
$(\frac{1}{2}, \frac{1}{2}, \pm \frac{1}{2})$	$(\frac{1}{2}, \frac{1}{2})$
$(\frac{3}{2}, \frac{1}{2}, \pm \frac{1}{2})$	$(\frac{1}{2}, \frac{1}{2})(\frac{1}{2}, \frac{3}{2})$
$(\frac{3}{2}, \frac{3}{2}, \pm \frac{3}{2})$	$(\frac{1}{2}, \frac{1}{2})(\frac{3}{2}, \frac{3}{2})$
$(\frac{5}{2}, \frac{3}{2}, \pm \frac{1}{2})$	$(\frac{1}{2}, \frac{1}{2})(\frac{1}{2}, \frac{3}{2})(\frac{3}{2}, \frac{3}{2})$

Fig. 25a shows the eigen-values of the unperturbed $\mathcal{H}_{\sigma\tau}$, fig. 25b the removal of degeneracy by spin-dependent forces and fig. 25c the effect of Coulomb forces regarded as a perturbation (114).

The $(2J + 1)$ degeneracy of *all* nuclear states is *not* included in references to the degree of degeneracy.

Considerations based on the approximation $\mathcal{H}_{\sigma\tau}$ to the nuclear Hamiltonian yield, primarily, information regarding the patterns of the energy levels. In order to predict actual level positions some knowledge is required of the relative magnitudes of the factors causing the splitting (Coulomb and spin-dependent) and the main symmetrical forces of $\mathcal{H}_{\sigma\tau}$. These factors can be estimated more reliably from a comparison of the binding energies of the ground states of isobaric nuclei than from the less complete and less reliable data about the energy levels of a single nucleus. Thus the Coulomb perturbation can be estimated by comparing the energies of the isobaric pairs $T_\zeta = \pm \frac{1}{2}$ (assuming they belong to the same T-multiplet), e.g. (C^{13}, N^{13}), (N^{15}, O^{15}). An estimate of the spin-dependent splitting can be obtained from the masses of light isobaric nuclei with A even but not a multiple of 4. For these nuclei

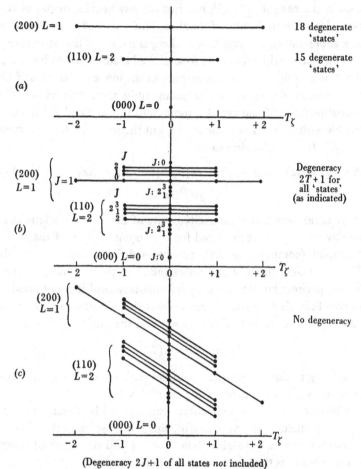

(Degeneracy $2J+1$ of all states *not* included)
Fig. 25. Typical super-multiplets for nuclei of even mass.
(Wigner and Feenberg, *Rep. Progr. Phys.* **8**, 274, 1942.)

the lowest super-multiplet is $(1, 0, 0)$ which comprises the T-multiplets $T = 1$, $S = 0$ and $T = 0$, $S = 1$. These two T-multiplets split under the influence of spin-dependent forces, which can be estimated by taking the difference between the mean binding energies of the two nuclei $T_\zeta = \pm 1$ and the energy of the single stable nucleus $T_\zeta = 0$.* The data for such a $(1, 0, 0)$ super-multiplet $(Be^{10} - B^{10} - C^{10})$ are shown in fig. 24 b.

In this case the Coulomb energy is insufficient to reduce the mass of $Be^{10}(T_\zeta = 1)$ to below that of $B^{10}(T_\zeta = 0)$,† a situation which is also true in the case of N^{14}, C^{14}, but not for any heavier nuclei of this type. If the ground states of the three nuclei with $A = 10$ all belong to a super-multiplet with $L = 0$, then, since $L - S$ fine structure is absent, one would expect the first excited state of B^{10} to belong to the $T = 1$ multiplet. The occurrence of the low excited state of B^{10} with energy 0.6 MeV. is not compatible with this view. The probable value of the angular momentum $(J = 1)$ of B^{10} is compatible with $L = 0$, 1 or 2 and $S = 1$, but theoretically $L = 0$ is most likely.‡ In the following cases

$$A = 6(\text{Li}, \text{He}); \quad 14(\text{N}, \text{C}); \quad 18(\text{F}, \text{O}); \quad 22(\text{Na}, \text{Ne}); \quad 26(\text{Al}, \text{Mg});$$
$$30(\text{P}, \text{Si}),$$

there is no evidence of an excited state of the $T = 0$ nucleus at an energy less than that expected for the 'point' $T_\zeta = 0$ of the $T = 1$ multiplet (compare fig. 23), as obtained from the mass of the $T_\zeta = +1$ stable nucleus and an estimate of the Coulomb energy. The experimental data are only fragmentary, and it is not possible to conclude that no such states really exist. In general, we might also expect the nucleus $T = 0$, $S = 1$ and the nuclei

$$T = 1, \quad S = 0 (T_\zeta = \pm 1)$$

belonging to the same super-multiplet with $L \neq 0$ to show different structures for the spectra of low-lying levels.

Comparison of experimental material with theory for the $T_\zeta = \pm \frac{1}{2}$ nuclei is not very useful, since in most cases the $T_\zeta = -\frac{1}{2}$ nucleus is much less stable than $T_\zeta = +\frac{1}{2}$ and the range of bound nuclear states is therefore much smaller.

* If both nuclei $T_\zeta = \pm 1$ are not known, the Coulomb energy can be estimated and the spin-dependent forces obtained from a comparison of the nucleus $T_\zeta = +1$ alone with $T_\zeta = 0$.

† The *binding energy* of Be^{10} is in fact greater than that of B^{10}, but the difference in neutron and proton mass makes the *mass* of Be^{10} greater than that of B^{10}.

‡ Recent experiments show the angular momentum of B^{10} in the ground state to be 3 (Gordy, Ring and Burg, *Phys. Rev.* **74**, 1191 (1948)), and also that of Na^{22} to be 3 (Davis, *Phys. Rev.* **74**, 1193 (1948)).

In two cases where the ranges of possible bound states are comparable (C^{11}, B^{11}), (O^{15}, N^{15}), the experimental evidence suggests roughly equal energies for the first excited level.

Apart from any uncertainty in the fundamental proposition on which the theory of super-multiplets is based (i.e. the assumption that $\mathscr{H}_{\sigma\tau}$ represents a valid first approximation), the range of application of this approach is limited in several respects. For large excitation energy a particular nuclear state will be a mixture of eigen-functions of $\mathscr{H}_{\sigma\tau}$ (all having the same J and T_ζ), the amount

TABLE 4

A		5		7		9	
T_ζ		$+\frac{1}{2}$	$-\frac{1}{2}$	$+\frac{1}{2}$	$-\frac{1}{2}$	$+\frac{1}{2}$	$-\frac{1}{2}$
Nucleus		He	Li	Li	Be	Be	B
First excited state (MeV.)		27·5*	26·7*	0·46	(?)	(?)	Unstable
Limit of bound states (MeV.)		Unstable		2·6	1·7	1·6	Unstable

A		11		13		15	
T_ζ		$+\frac{1}{2}$	$-\frac{1}{2}$	$+\frac{1}{2}$	$-\frac{1}{2}$	$+\frac{1}{2}$	$-\frac{1}{2}$
Nucleus		B	C	C	N	N	O
First excited state (MeV.)		2·1	2·2	0·8	(?)	5·3	4·0
Limit of bound states (MeV.)		8·5	7·2	4·9	1·9	10·6	7·2

A		17		19		21	
T_ζ		$+\frac{1}{2}$	$-\frac{1}{2}$	$+\frac{1}{2}$	$-\frac{1}{2}$	$+\frac{1}{2}$	$-\frac{1}{2}$
Nucleus		O	F	F	Ne	Ne	Na
First excited state (MeV.)		0·9	(?)	(?)	(?)	0·35	?
Limit of bound states (MeV.)		4·1	0·6	4·0	3·5	7·3	3·2

* Binding energy of virtual level.

of mixing being largest when the differences between the eigen-values of $\mathscr{H}_{\sigma\tau}$ are comparable with, or smaller than, the perturbation energies associated with spin and charge dependence of nuclear forces, and consequently most serious at high energies where the level density is large. The most stable states of heavy nuclei will not belong to super-multiplets with the smallest values of (P, P', P'') (number of neutrons much greater than that of protons, i.e. $T_\zeta \gg 0$) on account of the large Coulomb energy; and since the complexity of the super-multiplet patterns, the density of eigen-values of $\mathscr{H}_{\sigma\tau}$ and the Coulomb perturbation all increase with nuclear size, the mixing of eigen-states of $\mathscr{H}_{\sigma\tau}$ will be progressively more serious for heavier nuclei. But for the *lowest* energy levels, particularly the most stable states of an isobaric set of nuclei, the treatment of the Coulomb energy as a perturbation may be more justified than might be expected from the size of the Coulomb perturbation energy alone, since although T is not in principle a good quantum number for large perturbations, T_ζ has a definite value for a particular nucleus. (Mixing of states with different T_ζ is, of course, negligible; witness the long lifetime of β-processes.) The most stable state will belong to the super-multiplet with the smallest partition numbers, P, P', P'', which contains a T-multiplet with $T = T_\zeta$. Super-multiplets with smaller partition numbers will not include a T-multiplet of sufficiently large T (since the maximum value of T is equal to P the largest of the partition numbers), and since the state in question is the most stable one, it will correspond to the *lowest* T-multiplet with $T = T_\zeta$ in the super-multiplet. Therefore the energy interval between it and the higher super-multiplets will be substantially larger than the *average* spacing of super-multiplets.

In addition to mixing eigen-states of $\mathcal{H}_{\sigma\tau}$ with different partition numbers (P, P', P''), the Coulomb and spin-dependent forces can also mix states corresponding to different eigen-values of $\mathcal{H}_{\sigma\tau}$ but with the same partition numbers (i.e. states belonging to similar super-multiplets but having different configurations). Wigner estimates that the spacing between states with the same partition number but different configurations may be much less than that between states corresponding to different partition numbers, so that there may be considerable mixing of states belonging to different eigen-values of $\mathcal{H}_{\sigma\tau}$, of a type which does *not* destroy the super-multiplet symmetry classification.

The effect of the Coulomb perturbation is, then, to destroy the classification of nuclear excited states in terms of partition quantum numbers, probably even for fairly light nuclei, but the application of the classification to the ground state can be usefully employed over a much wider range. It is, in fact, in exploring properties of the normal nuclear state, binding energy, angular momentum, magnetic moments, etc., that this approximation to the nuclear Hamiltonian is most successful (114).

The limitations imposed by spin-dependence of the nuclear forces cannot be estimated with such confidence, since the nature and magnitude of the forces themselves cannot be assessed as reliably as can Coulomb forces.

Spin-dependent forces are normally classified as either spin-exchange forces of the Heisenberg-Bartlett type [115] (which do not depend explicitly on the electric charge of the interacting nucleons), or magnetic interaction type forces ('spin-spin' and 'spin-orbit'). With forces of the first type present, total orbital angular momentum, and total spin-angular momentum remain good quantum numbers (as for the more general case of eigen-functions of $\mathscr{H}_{\sigma\tau}$), and the degeneracy due to different L-S combinations (different J) is not removed. Forces of the latter type separate states with the same L and S but different J; and L, S can only be regarded as approximate constants of motion. The ground state and first excited state of Li^7 with excitation energy 0·48 MeV. [116] are usually interpreted as belonging to a spin doublet of this type, being respectively $^2P_{\frac{3}{2}}$ and $^2P_{\frac{1}{2}}$ in the usual spectroscopic notation. Only those states with different L, S will be mixed that have the same J, so that Russell-Saunders type coupling may be applicable even when the level spacing is of the same order as, or smaller than, the L, S splitting, but in view of the magnitude of this splitting (cf. Li^7 case above) it is unlikely that such a simple coupling scheme will apply to heavy nuclei or even for lighter nuclei in any except the lowest states.

Finally, it must be emphasized that considerations based upon approximate symmetry properties of the nuclear Hamiltonian indicate only the pattern of the excited states to be expected, and such features as variation of the level spacing with symmetry properties of the nucleus ('odd-even' characteristics). To make predictions about the *absolute* values of the excitation energies of the nuclear states, some particular nuclear model must be adopted.

20. NUCLEAR MODELS

In attempting to interpret nuclear spectra, or other nuclear properties, in terms of a specific model, the study of the actual complicated nuclear system is replaced by an investigation of a much simpler system whose properties can be regarded as roughly similar to those of the nucleus. Forces between individual nucleons are replaced, at least in the first approximations, by parameters which are chosen empirically; differences between the model and the actual nucleus are treated, if at all, as (small) corrections in higher approximations. The models which have been used most extensively are:

(i) Independent-particle model ('Fermi gas').

(ii) Independent-particle model with higher approximations ('Hartree approximation').

(iii) α-Particle model.

(iv) Rigid-body model.

(v) Liquid-drop model.

We shall consider briefly the methods, degree of success and particularly the scope of these models.

20·1. Independent-particle model

In this model each nucleon is assumed to move in a fixed potential, which represents the average effect of all other nucleons. Interaction between *individual* nucleons is otherwise neglected. The fixed potential may be chosen differently for neutrons and protons. The total nuclear wave function is then simply the product of the wave functions of the individual nucleons, and the energy the sum of the energy eigen-values corresponding to the individual particle eigen-functions. In order to represent, with reasonable accuracy and without undue mathematical complication, the average effect of the short-range nuclear forces, simple functions have been used for the equivalent fixed field, e.g. the oscillator potential ($V = -V_0 + Cr^2$), infinite potential well ($V = -\infty, r < R; V = 0, r > R$), the finite potential well ($V = -V_0, r < R; V = 0, r > R$), and the Gaussian potential well ($V = -De^{-r^2/R^2}$), where r is the distance from the centre of the nucleus and R some characteristic length of nuclear dimensions. All these potentials can be arranged, by suitable adjustment of the appropriate parameters, to give similar quantitative results for the single-particle energy eigen-values of the lowest energy levels, although they vary considerably in the degree of degeneracy of these levels. Thus the wave function of the nucleon in the oscillator potential can be characterized by three quantum numbers n_x, n_y, n_z corresponding to the three normal vibrations (in the x, y, z directions) and the energy is $E = -V_0 + \frac{1}{2}h\nu(N + \frac{3}{2})$, $N = \sum_{xyz} n$, and ν the classical oscillation frequency. Representing the wave functions as products of a radial function and a spherical harmonic (corresponding to different orbital and azimuthal angular momentum quantum numbers), for comparison with other central fields, involves taking linear combinations of eigen-functions corresponding to different sets of n's, but since only the total N determines the energy and the levels are equally spaced it is clear that there will be large degeneracy—actually $N(N+1)$-fold degeneracy allowing a factor 2 for spin degeneracy (117). Only the degeneracy with respect to spin and azimuthal quantum number occurs in the case of the rectangular potential well (finite or infinite). Both finite and infinite potential wells give similar level spacing of energy levels for the lower levels, provided the finite potential well is deep enough to accommodate a number of bound levels much larger than the number being compared. The average level spacing is smaller for the finite than for the infinite rectangular well. The level positions of the first few levels (in energy units of $\hbar^2/mR^2$) relative to the lowest level of the infinite potential well are given in table 5.

TABLE 5

Energy ($\hbar^2/mR^2$) ...	0	5·2	11·7	14·8	19·5	24·9	28·6	36·4
Level ...	1s	2p	3d	2s	4f	3p	5g	4d
l ...	0	1	2	0	3	1	4	2
No. of states	2	6	10	2	14	6	18	10
Complete shell	2	8	18	20	34	40	58	68

Each level is characterized by its energy, its position in a sequence of levels with a particular orbital angular momentum, l (the lowest level of the sequence being designated by the principal quantum number $(l+1)$), the degeneracy of the level, $2(2l+1)$, and the number of particles, neutrons or protons that are required to fill the level and all the levels of lower energy (complete shells).

The energy intervals between successive levels show pronounced fluctuations; notice the relatively large separation between 2s and 3p and between 5g and 4d levels. Interpreted literally this model predicts systematic variations in the energy of nuclei with number of nucleons, appearing as minima in the mass curve at the points at which complete shells are formed (e.g. $Z = 20$, $N = 20$ and $Z = 40$, $N = 58$, i.e. completion of 2s shell and of 5g shell for neutrons respectively). If we take into account the interaction between individual nucleons however, the degeneracy associated with a particular distribution of nucleons amongst single-particle levels (i.e. a particular 'configuration' so-called) will be removed, and except in the case of complete or almost complete shells, a single configuration will correspond to a very large number of actual levels, concerning the spacing of which the independent-particle model reveals nothing. Thus the zero-order approximation may be useful in interpreting the general variation of nuclear mass with size for light nuclei; it might also be of some value in the interpretation of the variation of the average level density with number of nucleons if more experimental data were available, but it can hardly be used as a guide to the interpretation of details of nuclear spectra.

20·2. Hartree approximation [118]

Using the wave functions for a single particle in a fixed potential well, a wave function for the whole nucleus having the symmetry properties required by the exclusion principle can be constructed, viz.

$$\Psi \text{ (nucleus)} = \begin{vmatrix} \psi_1(p_1) & \psi_2(p_2) & \cdots & \psi_z(p_1) \\ \psi_1(p_2) & \psi_2(p_2) & \cdots & \psi_z(p_2) \\ \cdots & \cdots & \cdots & \cdots \\ \psi_1(p_z) & \cdots & \cdots & \psi_z(p_z) \end{vmatrix} \times \begin{vmatrix} \phi_1(n_1) & \cdots & \phi_N(n_1) \\ \cdots & \cdots & \cdots \\ \cdots & \cdots & \cdots \\ \phi_1(n_N) & \cdots & \phi_N(n_N) \end{vmatrix}. \quad (20 \cdot 1)$$

$\psi_1(p) \ldots \psi_z(p)$ are single-particle wave functions for the Z protons and $\phi_1(n) \ldots \phi_N(n)$ the wave functions for the N neutrons. The total energy can be derived from the expression

$$E = \int \Psi^* \left\{ V - \hbar^2/2m \left[\sum_{i=1}^{z} \nabla_{p_i}^2 + \sum_{i=1}^{N} \nabla_{n_i}^2 \right] \right\} \Psi \, d\tau \Big/ \int \int |\Psi|^2 \, d\tau, \quad (20 \cdot 2)$$

where V, the potential energy, is expressed as the sum of terms representing the interactions between pairs of particles,

$$V = \sum_{i=1}^{z} \sum_{k=1}^{N} U_{ik}(p_i n_k) + \frac{1}{2} \sum_{i=1}^{z} \sum_{k \neq i} P_{ik}(p_i p_k) + \frac{1}{2} \sum_{i=1}^{N} \sum_{k \neq i} N_{ik}(n_i n_k).$$
$$(20 \cdot 3)$$

It should be noticed that in this approximation the 'potential well' is not an essential feature, it serves merely to provide a set of individual particle wave functions and hence a total wave function. The energy can be computed from (20·2) and (20·3) provided some definite assumption is made about the forces between individual nucleons. The success and practicability of the method depends largely on the choice of the initial independent-particle wave functions. For very rough calculations plane-wave functions have been used to estimate binding energies, but this procedure is certainly not refined enough to give the properties of excited states.

Somewhat more detailed results are obtained if the wave functions of the rectangular potential well are used and some specific assumptions are made about the relevant combinations of possible single-particle wave functions. Assuming Russell-Saunders coupling, Bethe and Bacher have computed the energy of the lowest states of Li^6, which on the single-particle model comprises two protons in the $1s$ states, one proton in a $2p$ state and the same configurations for the three neutrons. The assumption of Russell-Saunders coupling means that of the many total nuclear wave functions (all degenerate in the zero-order approximation*) that belong to this configuration of nucleons, only those particular linear combinations that correspond to definite total orbital angular momentum L are to be evaluated, i.e. S, P, D states. The calculation is relatively simple in this case since with only one neutron and one proton in the $2p$ shell, the choice of linear combinations is unrestricted by the exclusion principle and the spin can be entirely neglected in this part of the calculation. The levels are found to be S, D, P in order of increasing energy, and all the levels can be either singlet or triplet (spin zero or unity), since there is no correlation between permissible total orbital and total spin-angular momentum in this simple case. The next higher approximation can be obtained if some assumption is made about spin-dependent forces. If these are assumed to cause alinement of the spins of the neutron and proton in the p state with parallel spins, then the lowest state of Li^6 will be triplet: $^3S_1$. There should be an excited state: $^1S_0$. This pair of states is not analogous to the Li^7 doublet, since in the present case the $^3S_1$ and $^1S_0$ states are split by Heisenberg type forces and not

---

* These correspond to the various possible z-components of angular momentum for the p-proton and p-neutron and to different spin functions.

as in the Li^7 case by L, S interaction. Comparison with the general properties of the (1, 0, 0) super-multiplet, and in particular with the binding energy of He^6 which should belong to the same T-multiplet as the $^1S_0$ state of Li^6, indicates that this latter state should have an excitation energy of several MeV. (The dissociation energy of Li^6 is only 3·7 MeV.)*

Calculations of this type have been extended to more general cases, where the restrictions of the exclusion principle are important in the evaluation of the correct linear combination of wave functions, by Wigner and Feenberg[120] and Feenberg and Phillips[119]. Wigner and Feenberg neglect Heisenberg spin-dependent forces, so that both total proton spin and total neutron spin are separately constants of motion; and for the total nuclear wave function all combinations of each term in the proton configuration with each term in the neutron configuration are taken which give a particular total orbital momentum (S, P, D, ... states) for the whole nucleus. The final state (or more precisely group of states degenerate in this approximation) is designated by the separate multiplicities (i.e. spins) of the neutrons and protons, e.g. $^{1,2}S$, $^{2,4}D$, etc. Calculations based on the assumption of Majorana forces only (similar results are obtained with a mixture of Majorana and Wigner forces) result in values shown in table 6 for the excitation energies of the few lowest states of some light nuclei. In this approximation two nuclei have the same level schemes if the occupied levels (corresponding to single-particle wave functions) in one nucleus correspond to the gaps in the incomplete shell of the other, e.g. N^{14} has one proton and one neutron missing in the $2p$ shell, Li^6 has one proton and one neutron only in this shell.

In table 6 the sets of terms for any one group of similar nuclei (e.g. Li^7, C^{13}, N^{13}) all correspond to the same configuration, i.e. in the zero-order approximation of no interaction between individual nucleons they would be degenerate. Since these same independent-particle wave functions are used in the actual approximation, the level separations given in the table arise from differences in *potential energy* when the different linear combinations of wave functions are used in evaluating the internucleon interactions.

* Feenberg and Phillips[119] estimate this singlet-triplet splitting of Li^6 to be 3·1 MeV. by using the singlet-triplet splitting of the deuteron as a measure of the strength of the Heisenberg forces.

The next higher approximation consists, as in·the case of Li⁶, of taking into account Heisenberg type forces, which will tend to reduce the energy of those states with proton and neutron spins parallel. Thus the ^{22}S terms of Li⁶ and B¹⁰ are resolved into $^3S_1$ and $^1S_0$, and the ^{22}P terms of Li⁸ and B¹² into ^3P and ^1P terms. This type of splitting will occur only if the separate multiplicities of neutrons and protons are both greater than unity. Finally, L–S interaction

TABLE 6

Nucleus	Term	Excitation energy (MeV.)	Nucleus	Term	Excitation energy (MeV.)
Li⁶ and N¹⁴	^{22}S	0	Be¹⁰ and C¹⁰	^{11}S	0
	^{22}D	2·2		^{11}D	1·2
	^{22}P	3·6		^{11}D	1·6
Li⁷, C¹³ and N¹³	^{12}P	0		^{11}F	3·0
	^{12}F	3·1		^{11}G	5·8
Li⁸ and B¹²	^{22}P	0	B¹⁰ *	^{22}S	0
	^{22}D	1·6		^{22}D	2·0
	^{22}P	2·9		^{22}D	3·3
Be⁸ and C¹²	^{11}S	0		^{22}F	4·2
	^{11}D	1·9		^{22}G	6·5
	^{11}D	6·3	C¹⁴ and O¹⁴	^{11}S	0
Be⁹, B⁹, C¹¹ and B¹¹	^{12}P	0		^{11}D	1·4
	^{12}D	1·2	N¹⁵ and O¹⁵	^{12}P	0
	^{12}F	3·1	O¹⁶	^{11}S	0
	^{12}G	5·6			

* The theory is insufficiently refined, apparently, to predict the correct order for the S and D terms. (Cf. footnote, 19·1, p. 109.)

gives rise to a 'fine-structure', in all cases, other than S terms, for which the *total* spin is not zero. Thus in successive approximations the term ^{22}P in Li⁸ splits into ^3P and ^1P (Heisenberg splitting) and $^3P_1$, $^3P_0$, $^3P_2$ and $^1P_1$ (L–S splitting). For the lowest state, ^{12}P, of N¹⁵ and O¹⁵ there will be no Heisenberg splitting but L–S splitting into $^2P_{\frac{3}{2}}$ and $^2P_{\frac{1}{2}}$ for which the separation should be of the same order of magnitude as for the doublet of Li⁷ (i.e. about 0·5 MeV.), and the $^2P_{\frac{3}{2}}$ state should have lower energy. There is no experimental confirmation of this 'fine-structure' of the ground states of O¹⁵ and N¹⁵. Results similar to those above are obtained by

Feenberg and Phillips [119] using a more elegant method based on the fact that all total nuclear wave functions corresponding to a particular configuration will be eigen-functions of $\mathcal{H}_{\sigma\tau}$ belonging to the same energy eigen-value (provided charge and spin independence is taken for granted in defining the individual particle wave functions, i.e. the same auxiliary potential well is employed for protons and neutrons and no explicit dependence on spin). The symmetry properties of $\mathcal{H}_{\sigma\tau}$ enable the matrix elements for the energy (using a generalized interaction between nucleons which include both spin-dependent and spin-independent forces) for linear combinations of wave functions corresponding to particular values of total orbital momentum to be established more readily. In this procedure the introduction of spin-dependence as a separate step in the successive approximation is eliminated.

In all the calculations, in order to obtain numerical estimate for level positions, it is necessary to make some assumptions about the relative strengths of the spin-dependent and spin-independent forces, and in addition to choose an explicit value for the radius and depth of the auxiliary potential well (i.e. to determine the single-particle wave functions). Values for these quantities are chosen which give the correct binding energies for the two- and three-nucleon systems (H_1^2, He^3, H^3) and the correct singlet-triplet splitting for the deuteron [121]. Most of these calculations are at least as limited in range of application by the factors mentioned in the discussion of the super-multiplet classification as is this classification itself, since the two approaches usually involve similar assumptions as to the relative magnitudes of the different types of forces used in the successive approximations. The assumptions particular to the independent-particle model and its successively higher approximations, which enable actual estimates of level positions to be made, are certainly not sufficiently justifiable in other than very light nuclei ($Z < 10$) to give reliable prediction about level positions.

20·3. α-particle model

The total binding energy of an α-particle is 28·2 MeV., i.e. 7 MeV. per nucleon, which is about four-fifths of the mean binding energy per nucleon for most nuclei, and an even larger fraction for most light nuclei ($A < 30$). This fact suggests an approximation to the nucleus by a model in which, except for a residue of odd neutrons and protons, nucleons are grouped together inside the nucleus into α-particles, and the interaction between these α-particles is sufficiently small compared with their binding energy for the time spent by the nucleus in other configurations to be negligible. This model, which is analogous to usual models of polyatomic molecules, has been investigated extensively, particularly in the case of nuclei which can be considered as composed entirely of α-particles (Be[8], C[12], etc.). The conditions under which the α-particle model is likely to be valid have been discussed fully by Wheeler (122), who gives as the two main requirements: (a) that the wave-length of any relevant disturbance in the nucleus shall be much larger than the size of the α-particle group, (b) that the period of vibration of any nuclear motion shall be short compared with the time taken for a nucleon of an α-particle to interchange positions with a particle in its surroundings ('diffusion time'). Both conditions indicate that the model will be applicable only to low-excitation energies, since the wave-length of a disturbance decreases with energy, and the diffusion time will decrease rapidly as the excitation energy approaches in magnitude the energy of the potential barrier (of order 10 MeV.) which inhibits the excursion of a constituent nucleon of an α-particle into its surroundings. The predominant features of the α-particle model are the additional special symmetry and the effective reduction in the number of degrees of freedom of the system, both factors, but especially the latter, resulting in a much larger level spacing than in the contrasting independent-particle model.

The mutual potential energy of two α-particles is assumed to be of the form shown in fig. 26 (123). At short distances ($r < r_1$) there is strong repulsion due to the operation of the exclusion principle (with respect to individual constituent nucleons), at intermediate distances ($r_1 < r < r_2$) there is a polarization attractive force (Van der Waals type of force). At large distances ($r > r_2$) there remains only the Coulomb repulsion. For small excitation energy, the separation between two α-particles will be approximately r_1, and for small amplitude 'vibrations' about this mean position the potential can be replaced by an 'oscillator' potential. In general two main types of motion (not completely independent) have to be considered: (a) rotation of the nucleus as a whole, (b) small (approximately simple harmonic) vibrations. If the coupling between the rotational

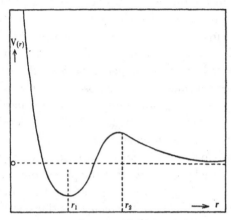

Fig. 26. Interaction potential for two α-particles.

and vibrational modes is neglected, the excitation energies in the case of Be^8 (dumbell), C^{12} (equilateral triangle) and O^{16} (tetrahedron) are (125)

Be^8: $E_{\text{rot.}} = (\hbar^2/2I)J(J+1)$ $(J = 0, 2, 4, ...)$;

C^{12}: $E = E_{\text{rot.}} + E_{\text{vib.}} = (\hbar^2/2I)\{J(J+1) - \tfrac{1}{2}K^2\} + n_1 h\nu_1 + n_2 h\nu_2$

$(|J| > |K|;$ $K = \pm 3m;$ $J = 0, 1, 2, ...;$ $n_1, n_2 = 0, 1, 2, 3, ...;$

$K = 0, J$ is even, m is integral);

O^{16}: $E = E_{\text{rot.}} + E_{\text{vib.}} = (\hbar^2/4I)\{J(J+1)\} + n_1 h\nu'_1 + n_2 h\nu'_2$

$+ n_3 h\nu'_3 \pm \Delta E$

$(J = 0, 3, 4, 6, 7, ...;$ $n_1, n_2, n_3 = 0, 1, 2, ...).$

In these expressions J, K, n_1, n_2, n_3 are total angular momentum, angular momentum along axis perpendicular to plane of figure (C^{12}), and vibration quantum numbers respectively. They are restricted (as indicated) by the geometrical symmetry of the nuclei. I is the moment of inertia of the Be^8 nucleus (assuming that the separation of a pair of α-particles is the same in all these nuclei); ν_1, ν_2 (for C^{12}) are the frequencies associated with normal modes of oscillation in which there is dilation of the system as a whole, and in which there is a change in the distance between one pair of

α-particles only; ν_1', ν_2', ν_3' (for O^{16}) correspond similarly to dilation, turning of the axis joining one pair of α-particles with respect to that joining the other pair, and change of distance between one α-particle pair without twisting; ΔE (for O^{16}) is a small-energy term, depending on J and n's, associated with the reflexion of the apex of the tetrahedron in the base ('inversion'). ν_1, ν_2, ν_1', ν_2', ν_3' are related to each other simply by geometrical considerations if the forces between α-particles are assumed the same in all cases. Quantitative predictions about level positions result from making estimates of I and the strength of the oscillator potential. I may be determined by judicious guesswork, by comparison of the predicted and experimental levels in a particular case (e.g. Be^8 level at about 3 MeV.) or by comparison of the α-particle and independent-particle models. A more satisfying procedure is to base a quantitative estimate on an analysis of the experimentally observed α-scattering. Two parameters, one for vibration and one for rotation, suffice to fix the positions of the lowest levels of all three nuclei. These are

Be^8:   S, D, G, ...   (i.e. J = 0, 2, 4). Vibrational states are ignored since in view of the small binding energy of Be^8 these would be well in the continuum.

C^{12}:   (0000), (2000), (3300), ..., (J, K, n_1, n_2).

O^{16}:   (0000), (3000), (4000), (0100), (2010), ..., (J, n_1, n_2, n_3)
(order of levels uncertain).

If the D state of Be^8 is identified with the observed level of energy 3 MeV., then $\hbar^2/2I$ = 0·5 MeV., and the first two rotational levels of C^{12} should have excitation energies of 3·0 and 3·75 MeV. The lowest observed levels of C^{12} are a doubtful level at 3·0 MeV. (Be^9 (α, n) C^{12}), and a well-established level at 4·35 MeV. (N^{14} (d, α), N^{15} (p, α), B^{11} (d, n) and Be^9 (α, n) reactions). Rotational level of O^{16} should similarly occur at 3·0 MeV. (J = 3) and 5·0 MeV. (J = 4). Experimentally the lowest excited level of O^{16} is at 6·0 MeV. [126], and this is an $S(J$ = 0) level, since the transition to the ground state occurs by emission of electron-positron pairs rather than γ-radiation [128]. If a level with J = 3 (odd parity) were located at about 3 MeV., the electric-octupole transitions to this level would be possible in strong competition with pair emission, but no such

radiation is observed.* Another level of O^{16} is well established at
6·2 MeV., and for this level γ-transitions to the ground state occur.
If it is to be associated with the (3000) level of the α-particle model,
then a rather small radius of O^{16} (the value of $\hbar^2/2I$ used above
corresponds to a radius of O^{16} of 3×10^{-13} cm.) must be assumed.
Alternatively, the shift of the level from the expected position might
be interpreted as an 'exchange' effect discussed by Teller and
Wheeler[128]. The value of ν_1' obtained by associating the 6·0 MeV.
level with the (0100) level of the α-particle model (i.e. symmetrical
dilation) is in reasonable accord with the known characteristics of
the interaction between two α-particles. Extensions of the α-particle
model to light nuclei of the $4k \pm 1$ type have been considered by
Hafstad and Teller[124]. As might be expected this model predicts
much larger level spacing for low-excitation energies than the
independent-particle model.

All α-particle models represent, of course, a very special case of
states with the lowest possible partition numbers (P, P', P''); e.g.
for $4k$ $(Z = N)$ type nuclei all states of the α-particle model are of
the (0, 0, 0) type in Wigner's classification. There is no sharp line
of demarcation between the α-particle model and the independent-
particle model, although they represent opposite extremes of a more
general treatment. The wave function of the whole nucleus can be
put into a form showing the association of the nucleons into any
particular groupings, and although in general an actual nuclear
state must be regarded as a superposition of states corresponding to
all possible groupings, approximately correct predictions can be
obtained by considering one grouping (e.g. the α-particle grouping
corresponding to lowest partition numbers) as predominant[122].

20·4. Rigid-body model[129]

The separation of the excitation energy of a system into rotational
and vibrational energies is equivalent to treating the system as
perfectly rigid as far as rotations are concerned (cf. the similar
approximation in treatment of molecular rotation[130]), and the
validity of this treatment depends on the relative magnitude of

---

* The half-life for an octupole γ-transition with energy 3 MeV. would be
about 10^{-12} sec. Pair emission occurs with half-life less than 10^{-6} sec. but
probably much larger than 10^{-12} sec. (unpublished results).

rotational and vibrational level spacings. If one disregards, for the moment, any limitations on the angular momentum J imposed by symmetry properties, then the rotational states of a rigid spherical nucleus of mass Am and radius R would have energies $(5\hbar^2/4AmR^2)$. $J(J+1)$, i.e. the first rotational level, would be about 10^4 eV. for $A = 200$, 3×10^4 eV. for $A = 100$, 9×10^4 eV. for $A = 50$ and $1 \cdot 2 \times 10^6$ eV. for $A = 10$. These values are, on the average, rather smaller than the observed level spacing, but the experimental results at present available indicate a wide variation in the position of the first excited level for nuclei of comparable mass. In particular, for nuclei of the even Z-even N type, the lowest observed excited level is in nearly all cases higher than is indicated by the rigid rotator model above, e.g. Mg^{24}, $1 \cdot 38$; Si^{28}, $1 \cdot 8$; Ti^{48}, $1 \cdot 0$; Cr^{52}, $1 \cdot 46$; Cr^{54}, $0 \cdot 835$; Fe^{56}, $0 \cdot 845$; Ge^{74}, $0 \cdot 58$; Se^{76}, $0 \cdot 83$; Kr^{82}, $0 \cdot 547$; Mo^{94}, $0 \cdot 4$; Pd^{108}, $0 \cdot 65$; Sn^{116}, $0 \cdot 43$; Te^{122}, $0 \cdot 96$; Te^{124}, $0 \cdot 6$; Xe^{126}, $0 \cdot 5$; Xe^{128}, $0 \cdot 43$; Xe^{130}, $0 \cdot 54$; Ba^{134}, $0 \cdot 79$; W^{182}, $0 \cdot 07$; Pb^{206}, $0 \cdot 19$ MeV. (but Se^{78}, $0 \cdot 046$; Cd^{112}, $0 \cdot 17$ ($0 \cdot 25$?) MeV.).

These nuclei all have zero angular momentum in the ground state, so that if a simple interpretation in terms of rigid rotation is attempted it can only be made to fit by assuming that these levels all correspond to large rotational angular momenta, $J \gtrsim 3$. The evidence is against such an interpretation since: (a) the I.C.C. of γ-transitions from such levels does not generally (where measured) suggest a value of J exceeding 2; (b) the excitation of the levels is usually through β-decay, and in many cases excitation in such a manner would be improbable if J were so large; (c) the lifetime for γ-transitions is normally less than 10^{-6} sec. (coincidence measurements), whereas $J \sim 3$ or 4 would correspond to a slower transition in many cases; (d) in any case on the rigid-rotator model lower states of smaller angular momentum would be expected, and the non-observation of such levels can hardly be due to their excitation being improbable (cf. §5·1; β-selection rules).

The fact that for actual nuclei the internal ('vibrational') level spacing (as we must interpret the above examples) is not very much larger than the rotational level spacing means that rotational motion will cause a certain mixing of the vibrational states, i.e. the separation of vibrational and rotational degrees of freedom is not strictly applicable. This feature in itself might not be expected to change the rotational level spacing by an *order of magnitude*, but considerations of the physical symmetry of the nucleus (arising out of the identity of the nuclear constituents) may enable the absence of low rotational states to be explained (128).

It has been stated in connexion with the α-particle model treated as a system with independent rotational and vibrational motions, that only certain values of the rotational angular momentum are possible on account of the spatial symmetry of the α-particle models in particular cases. Consider, for example, the simple case of a two-dimensional rigid model for which a rotation of $\theta(= 2\pi/f)$ about the axis of symmetry in the third dimension is equivalent to an exchange of co-ordinates of each pair of neutrons and protons. Then the rotational wave function for such a model satisfies the relation

$$\psi(\theta) = \psi(\theta + 2\pi/f). \tag{20.4}$$

Since the dependence of the rotational wave function on θ is of the type $\cos J\theta$, where J is an integer (angular momentum quantum number), then J can only take the values f, $2f$, $3f$, etc., i.e. the higher the symmetry of the system the larger will be the spacing of the rotational levels. This suppression of the rotational levels for which $J \neq f$, $2f$, etc., is only strictly true if the system possesses the perfect symmetry indicated by (20.4), but if the nucleus is also executing internal vibrations, then this perfect symmetry may not be maintained, and rotational levels for which $J \neq f$, $2f$, etc., may then appear, but the lowest of such rotational levels will have a (large) excitation energy corresponding to excitation of a vibrational mode as well as rotation. Thus the extreme case (applicable to an ideally rigid nucleus) of suppression of rotational states is replaced by the apparent shift of some of the rotational states to much higher energy. The converse effect will affect the position of the lowest rotational states. If the nucleus does *not* possess exact symmetry in its lowest vibrational state (ground state), then there will be no restrictions on the values which the rotational angular momentum can assume. But since the kinetic energies of the nucleons in the lowest vibrational state are not in fact zero, the nucleus may be regarded as spending a certain fraction of time in a perfectly symmetrical state for which certain rotational states are impossible. If the period of the transitions between the actual and the ideal symmetrical configurations is of the same order of magnitude as (or smaller than) the time associated with rotation of the nucleus through an angle $2\pi/f$, then although states with angular momenta not equal to $2\pi/f$, etc., will not be suppressed completely (as they are for a rigid symmetrical body), the effect of the approximate symmetry will be to displace these levels substantially, and the nucleus will then behave similarly to a non-rigid system with a high degree of geometrical symmetry in the ground state. The magnitude of these level displacements will depend on the relative energies of the (approximately separate) vibrational and rotational energies and the closeness to exact symmetry.

Teller and Wheeler (128) estimate, by means of a rough semi-quantitative argument, that the displacements will be so large that the *first* rotational state for a heavy nucleus will have angular momentum $J \sim 1 \cdot 7 \times A^{\frac{1}{3}}$, but an individual nucleus might be expected to differ widely from this rough estimate. Experimentally, there is a systematic difference between the position and angular momentum of the lowest states of nuclei of the even-even type and those of the other nuclei. For example, the lowest states of some nuclei with odd Z, odd A, or both odd are: V_{23}^{49}, $0 \cdot 18$; Cu_{29}^{63}, $0 \cdot 28$; Co_{27}^{60}, $0 \cdot 056$; Zn_{30}^{67}, $0 \cdot 093$; Zn_{30}^{69}, $0 \cdot 44$; Ge_{31}^{71}, $0 \cdot 60$; As_{33}^{75}, $0 \cdot 04$; Br_{35}^{80}, $0 \cdot 037$; Kr_{36}^{83}, $0 \cdot 046$; Sr_{38}^{85}, $0 \cdot 17$; Tc_{43}^{99}, $0 \cdot 136$; Ru_{43}^{103}, $0 \cdot 04$; Cd_{48}^{115}, $0 \cdot 16$; In_{49}^{114}, $0 \cdot 19$; Te_{52}^{127}, $0 \cdot 086$; I_{53}^{133}, $0 \cdot 177$ MeV.

These energies are much smaller than the corresponding energies for even-even nuclei, and, in addition, the angular momentum difference between the first excited level and ground level is frequently greater than 3 for nuclei with odd Z or N. This marked difference between even-even and other nuclei can be explained as due to the larger displacement of the rotational levels of even-even nuclei on account of their higher spatial symmetry. It is plausible to assume that the rotational levels for small J (say $J \lesssim 4$) are displaced to energies greater than the first vibrational level, and the lowest rotational level $J \sim 4$, even if slightly displaced, would have energy ($\sim 0 \cdot 5$ MeV. for $A = 100$, $J = 5$) at least comparable with the vibrational energies. Thus for even-even nuclei the lowest level may well be a *vibrational* level with small angular momentum. For nuclei with odd Z or N, we may assume that the displacement of the rotational states is usually smaller on account of the less pronounced spatial symmetry, but is still sufficient to displace the few lowest rotational states (say $J = 1, 2$) to higher energies than those with larger J values, and the lowest observed level in this case will be a *rotational* level with $J \gtrsim 3$.* For $A = 100$, $J = 3$ corresponds to an excitation energy 0.18 MeV., which is in reasonable agreement with observed values (cf. also fig. 22).

20·5. Liquid-drop model

This model, first proposed by Bohr and Kalkar [131], represents an extreme simplification in a different direction from those of the previous models; namely, the assumption that the constituent particles of the nucleus are sufficiently numerous for their discreteness to be ignored and the nucleus to be treated as a continuum. In the liquid-drop model two distinct forms of energy are recognized: surface energy and volume energy, associated respectively with vibrations in which the surface changes without volume change, and vibrations in which, in general, both volume and surface area change. Quantitative estimates of the constant of proportionality in the ratio of surface energy to surface area, and of rate of change of energy with volume can be deduced from the known variations of nuclear binding energy with nuclear mass [132]. The lowest energy levels (apart from possible rotational levels) correspond to surface vibrations. Regarding the behaviour of the nucleus as the same as that of an ordinary classical liquid drop, then the first excited state will correspond to the fundamental mode of vibration (i.e. oscillation between ellipsoidal and spherical shape) with frequency $\omega_0 \sim (F/\rho R^3)^{\frac{1}{2}}$, where F is the surface energy per unit area and ρ, R the nuclear density and radius. Using the values $R = 1.5 \times 10^{-13} A^{\frac{1}{3}}$ cm. and the surface energy equal to $1.3 A^{\frac{2}{3}}$ MeV., we get for the energy of the lowest surface vibration, $\hbar \omega_0$, the value $15 A^{-\frac{1}{2}}$ MeV. (1 MeV. for $A = 200$, 5 MeV. for $A = 10$). It is perhaps surprising that this result is even of the correct order of magnitude, since the assumption that nuclear matter can be treated like an ordinary liquid would only be justified if the kinetic energy associated with the oscillatory motion were large compared with the zero-point energy. In fact, for low-excitation energy the zero-

---

* For nuclei having angular momentum J_0 in ground state the 'rotational' state may have angular momentum between $|J-J_0|$ and $|J+J_0|$.

point energy is much larger than the excitation energy (i.e. most of the nucleons are in the lowest states permitted by exclusion principle), so that, as has been pointed out frequently, the behaviour of the nucleus, if regarded as a liquid drop, will be more akin to that of a liquid at very low temperatures, so-called quantum liquids such as liquid He II.

21. STATISTICAL PROPERTIES OF MODELS

In most respects in which the nuclear models discussed above give explicit information about level positions, the results are only applicable, at best, to the few lowest levels (except perhaps in the case of very light nuclei). They are not sufficiently refined to be applied to nuclei with large excitation energies on account of the very complicated nature of the dynamical systems representing such excited nuclei. But, as in other branches of physics, this very complication can be turned to advantage if inquiry is limited to the statistical properties of nuclear level distributions. There is one outstanding condition that must be satisfied for such treatment to be applied successfully to any particular model, namely, that for the excitation energies concerned there must be a statistically large number of effective degrees of freedom. This implies not only a large number of constituent nucleons but also an excitation energy sufficiently large for the density of levels of the nucleus as a whole to be considered as a continuous function of energy. When these conditions are satisfied, the usual methods of statistical mechanics can be employed.

The 'temperature' T (in energy units) of a nucleus with excitation energy E can be defined by the expression

$$E = \sum_i g_i \epsilon_i \exp(-\epsilon_i/T) / \Sigma g_i \exp(-\epsilon_i/T), \qquad (21 \cdot 1)$$

where g_i, ϵ_i are the statistical weight (degeneracy) and energy of the ith level, and the summation is over all energy *levels* of the nucleus.

In the limiting case of large level density (21·1) can be written

$$E = \int \epsilon \rho(\epsilon) \exp(-\epsilon/T) d\epsilon \Big/ \int \rho(\epsilon) \exp(-\epsilon/T) d\epsilon, \qquad (21 \cdot 2)$$

where $\rho(\epsilon)$ is density of nuclear *states*.

T can be interpreted physically by considering an assembly of nuclei in statistical equilibrium with a dilute gas of nucleons contained in a finite enclosure. It is assumed that the nucleons can excite the nuclei by inelastic collision. Then if the energy of the nucleon gas is maintained at a value sufficient for the mean excitation energy of the nucleus to be E, the temperature of the nucleus, T, will be that characterizing the nucleon gas (mean kinetic energy of the nucleons $\frac{3}{2}T$) (cf. §2·2). If the usual requirements for the applicability of statistical arguments are satisfied, then characterization of the temperature will not only specify the average excitation energy but also results in most of the systems (nuclei) in equilibrium at temperature T, having energies very near to $E(T)$, i.e. the contribution to the integral in (21·2) comes only from a small range of energies $\delta\epsilon$ near the mean.

The entropy S of the nucleus can be defined by the relation

$$S = \log[\rho(\epsilon)\,\delta\epsilon], \qquad (21\cdot3)$$

since $\rho(\epsilon)\,\delta\epsilon$ represents the effective number of states for a given temperature. $\delta\epsilon$ varies only slowly with E, whereas $\rho(\epsilon)$ increases rapidly, so that approximately* $\rho(\epsilon) \propto e^{S}$. From detailed considerations of a particular system one usually derives E as a function of T, so that to obtain $\rho(E)$ we use the usual relation $dS/dE = 1/T$, so that

$$S = \int \frac{1}{T}\frac{dE}{dT}dT, \quad \rho(E) = \exp\left[\int \frac{1}{T}\frac{dE}{dT}dT\right]. \qquad (21\cdot4)$$

Different nuclear models give different dependence of E on T.

21·1. Independent-particle model

The independent-particle model is equivalent, if the number of particles is sufficiently large, to the well-known degenerate electron gas for which Fermi-Dirac statistics apply. If we assume that the size of the nucleus is large compared with the de Broglie wavelength of the nucleons in the higher independent-particle levels (i.e. at the top of the Fermi-Dirac distribution), then the density of energy levels of the individual nucleons can be assumed to be a continuous function of the energy of the nucleon. This approximation is justified only for large nuclei. Regarding the nucleus as composed of equal numbers of neutrons and protons, then since

* More precisely (133),
$$S = \log \rho + \tfrac{1}{2}\log\,(T^{2}2\pi\,dE/dt). \qquad (21\cdot4a)$$

each spatial wave function for an individual nucleon can represent four nucleons (two values of spin and isotopic spin), we can write for the energy of the whole nucleus with temperature T [134]

$$E = \tfrac{3}{5}A\epsilon_c + A(\pi^2 m T^2/2h^2)(16\pi V/3A)^{\frac{2}{3}}, \qquad (21\cdot5)$$

where

$$\epsilon_c = (h^2/8m)(3A/2\pi V)^{\frac{2}{3}}. \qquad (21\cdot6)$$

In this expression, which is applicable in the case of low temperatures ('extreme degeneracy'), A is the number of nucleons and V the nuclear volume. The constant term in $(21\cdot5)$ is of no significance for the thermodynamic treatment, and we can rewrite $(21\cdot5)$ as

$$E = \tfrac{1}{4}\pi^2(AT^2/\epsilon_c).$$

Using $(21\cdot4)$ we obtain

$$S = \tfrac{1}{2}\pi^2 AT/\epsilon_c = \pi(AE/\epsilon_c)^{\frac{1}{2}}, \qquad (21\cdot7)$$

which together with $(21\cdot4a)$ (footnote p. 134) gives

$$\exp S = 2\sqrt{2}\,A^{-\frac{1}{4}}\epsilon_c^{\frac{1}{4}}E^{\frac{3}{4}}\rho(E),$$

i.e.

$$D(E) = 1/\rho(E) = 2\sqrt{2}\,A^{-\frac{1}{4}}\epsilon_c^{\frac{1}{4}}E^{\frac{3}{4}}\exp\left[-\pi\sqrt{(AE/\epsilon_c)}\right]. \qquad (21\cdot8)$$

If we make the usual assumption that the nuclear volume is proportional to A, then ϵ_c is independent of the particular nucleus, and for $R = 1\cdot5 \times 10^{-13}A^{\frac{1}{3}}$ cm., $\epsilon_c = 21\cdot4$ MeV. The dominant factor in $(21\cdot8)$ is the term $\exp\left[-\pi\sqrt{(AE/\epsilon_c)}\right]$, which, with the above value for ϵ_c, gives a decrease of level spacing with A which is almost certainly too rapid. The numerical values obtained from $(21\cdot8)$ also indicate a level spacing smaller (e.g. for $E = 8$ MeV., i.e. about the binding energy of a neutron, $D(E) = 0\cdot03$ MeV. for $A = 100$; and for $E = 7$ MeV. $A = 200$, $D(E) \sim 0\cdot5 \times 10^{-3}$ MeV.) than is suggested by experiments with slow neutrons. Of the numerous assumptions and approximations inherent in this treatment, the major ones, together with refinements that have been attempted to improve the theory, are indicated below.

(i) The approximation that the number of nucleons, A, and hence also the effective number of degrees of freedom, is very large. This is the usual assumption made in a statistical-mechanical treatment, and for $A \sim 100$ and excitation energies ~ 5 MeV. should not lead to serious error.

(ii) The use of the ordinary Fermi-Dirac distribution function to compute the dependence of E on the nuclear temperature T.

This is only justified, as mentioned above, if the wave-length of the nucleons near the top of the individual particle-energy distribution is very small compared with nuclear dimensions. In view of the results quoted in §20·1, which indicate that for a nucleus with $A \sim 120(Z \sim 50, N \sim 70)$ all the nucleons can be accommodated in an infinitely deep potential without going to higher wave functions than $5g$, it is clear that the simple Fermi distribution will give only a rough order of magnitude for the level spacing, and, furthermore, will not exhibit any fluctuations in level density with the number of component nucleons, as a complete model might. From an inspection of the variation of nuclear binding energy with A one can obtain some evidence of shell structure in nuclei, which finds a ready, if only qualitative, explanation in terms of the independent-particle model. One might expect to find similar variations in level density. Margenau [135] has investigated the independent-particle model statistically, but using the actual discrete level spectrum for individual particles in an infinite potential well in place of the approximate, continuous distribution of single-particle energy levels which is implicit in (21·5). The assumption of large A, and hence a continuously varying level density for the *whole* nucleus, is still involved in the statistical treatment. Thus instead of the usual expression $8\pi V/h^3(2m^3\epsilon)^{\frac{1}{2}} d\epsilon$ for the number of single-particle levels in the energy interval ϵ, $\epsilon + \delta\epsilon$ (used in deriving 21·5), Margenau uses the actual single-particle level positions ϵ_i and degeneracies g_i to find the temperature equivalent to an excitation energy E, viz.

$$\Sigma \frac{g_i}{1 + \alpha \exp(\epsilon_i/T)} = A \quad \text{(number of nucleons)},$$

$$\Sigma \frac{g_i\epsilon_i}{1 + \alpha \exp(\epsilon_i/T)} = E \quad \text{(total excitation energy of nucleons)}.$$

These expressions can be used for any particular distribution of the levels (ϵ_i, g_i), and if values of α and T can be found which satisfy them, then from the relation between T and E, the level density can be obtained as a function of excitation energy as before. The results of this more refined treatment show that within the limitations of the independent-particle model, the variation of level spacing with A shows a marked fluctuation; e.g. near $A = 116$ (completion of $5g$ shell) and for excitation energies of 8 MeV. the level *spacing* is

5×10^4 times *greater* than that obtained by the rough method, but for $A \sim 100$ the level spacing is 50 times *smaller*. Such fluctuations in level density very probably occur, but their magnitude and location (with respect to magnitude of A) may be very different from the predictions of this 'zero-order' approximation when the effects of inequality of number of neutrons and protons and the interaction between nucleons are taken into account. Both factors will tend to blur out any sharp discontinuities observable when a level shell is completed.

(iii) The level density depends not only on A, but on N and Z separately. Increase of $(N - Z)$, with constant A, leads to an increase in the level density but not by a large factor (136).

(iv) The independent-particle model is likely to give too large a result for the level density, since it corresponds to the maximum number of effective degrees of freedom for the nuclear system. Bardeen (137) has shown that if the internuclear forces are predominantly exchange-type forces, then there will be a tendency for strong correlation of the relative positions of groups of (up to four) nucleons, thereby reducing the number of *effectively independent* particles. This factor is estimated as roughly equivalent to a reduction by a factor 2 in the number A, of independent nucleons (although in estimating the value of ϵ_c the full value A must be used in accordance with the exclusion principle). This will change the important exponential factor in (21·8) to $\exp[-\pi \sqrt{(AE/2\epsilon_c)}]$, so increasing the level spacing and also reducing the dependence on A and E. With this modification the values for level spacing become

$$A = 100, \quad E = 8 \text{ MeV.}: \quad D(E) = 10·6 \text{ eV.}$$
$$A = 200, \quad E = 7 \text{ MeV.}: \quad D(E) = 1·1 \times 10^{-2} \text{ eV.,}$$

values which are still, probably, smaller than actual level spacings.

(v) The density of levels as given by the simplified statistical treatment above is, strictly speaking, the density of states, and each level corresponds to $(2J + 1)$ degenerate states (J is total nuclear angular momentum). To obtain the actual level density one needs to know in addition the statistical distribution over J of the states. Bethe (133) has given a more refined statistical treatment of the independent-particle model in which the density of energy levels of particular J is evaluated as

$$D(E, J) = D(E, O)/(2J + 1),$$

where $D_0(E, O)$ is the level density for states of zero J given by the expression

$$D(E, O) = 1\cdot43\, A^2 E^2 \epsilon_c^{-1} \exp\left[-\pi\sqrt{(AE/2\epsilon_c)}\right], \qquad (21\cdot9)$$

in which the effective reduction in A by a factor 2 has been included (cf. (iv)).

(vi) In treating the single-particle model non-statistically (§ 20·1), the levels of the whole nucleus are given directly by the energy levels of the individual nucleons. The lowest levels correspond to excitation of one, or of a few, of the least firmly bound nucleons into the next unoccupied levels. Since the individual level spacing does not increase markedly with energy, the level density of the whole nucleus will not increase very rapidly. However, as the excitation energy increases, the degeneracy of individual levels increases, and the degeneracy of levels of the whole nucleus even more rapidly. When interaction between individual nucleons is taken into account, these highly degenerate levels are split in many separate levels (only the degeneracy of $2J+1$ per level remains), resulting in a level density which increases rapidly with excitation energy, i.e. the continuum of levels assumed in the statistical model with large A and high excitation. In actual nuclei (particularly for small A) we should expect some state of affairs intermediate between the small number of discrete degenerate levels and the continuum of non-degenerate levels, i.e. a discrete spectrum with no absolute degeneracy (except $2J+1$), but showing marked fluctuations in level density with varying excitation energy. A similar conclusion is reached through the approximations on which the super-multiplet classification is based.

21·2. Liquid-drop model

The application of the statistical treatment to the liquid-drop model follows a pattern similar to that of the previous case, with the difference that in this case the analogy is not to a Fermi-Dirac electron distribution, but approximately to the Debye theory of specific heats in an elastic continuum. For moderate excitation energies, of the order of the binding energy of a neutron or less, the most important modes of vibration of the whole nucleus will be surface vibrations at constant volume. For these, the number of normal modes with angular frequency between ω and $\omega+d\omega$ is

$\frac{4}{3}R^2(\rho/F)^{\frac{3}{2}}\omega^{\frac{1}{2}}d\omega$ (notation as in §20·5). The average energy per normal mode, for a nuclear temperature T, is $\hbar\omega/[\exp(\hbar\omega/kT)-1]$. Hence the total energy as a function of T is

$$E(T) = \tfrac{4}{3}R^2(\rho/F)^{\frac{3}{2}}\hbar\int_0^\infty \frac{\omega^{\frac{3}{2}}d\omega}{\exp(\hbar\omega/kT)-1}$$

$$= \tfrac{4}{3}R^2(\rho/F)^{\frac{3}{2}}\hbar(kT/\hbar)^{\frac{5}{2}}\int x^{\frac{3}{2}}/(e^x-1)\,dx. \qquad (21\cdot10)$$

Putting $R = r_0 A^{\frac{1}{3}}$, $k = 1$,

$$E(T) = 2\cdot26[3Amr_0^2/\zeta\hbar^2]^{\frac{3}{2}}T^{\frac{5}{2}},$$

where ζ is the constant in the relation: surface energy $= \zeta A^{\frac{2}{3}}$. Using the values $\zeta = 13\cdot0$ MeV., $r_0 = 1\cdot5 \times 10^{-13}$ cm., we get

$$E = 0\cdot125A^{\frac{3}{2}}T^{\frac{5}{2}}. \qquad (21\cdot11)$$

Proceeding as in the previous case, we obtain for the level spacing

$$D(E) = 8\cdot1 \times 10^5 E^{\frac{7}{4}}A^{\frac{1}{2}}\exp\left(-0\cdot72A^{\frac{3}{4}}E^{\frac{2}{5}}\right) \text{eV}. \qquad (21\cdot12)$$

(E, T are in MeV.)

Typical examples are

$$A = 100, \quad E = 8\,\text{MeV.}, \quad D(E) = 10^3\,\text{eV.}; \qquad A = 200,$$
$$E = 7\,\text{MeV.}, \quad D(E) = 300\,\text{eV.}$$

These values are not only much larger than those obtained from the independent-particle model but the decrease of level density with A is less rapid, probably in better accord with the actual variation of level density. The numerical values for $D(E)$ depend sensitively on the values of r_0 and ζ adopted.

Apart from the limitations of the liquid-drop model already pointed out there are additional, very questionable assumptions involved in the statistical treatment. The integration in $(21\cdot10)$ over the normal modes is valid only if the spectrum is effectively continuous for the frequencies involved, i.e. if the wave-lengths of the modes are *small* compared with nuclear dimensions. But in order to treat the nuclear medium as a continuum the wave-lengths of the vibrations must be *large* compared with internucleon separation, which, even for $A = 200$, is about one-sixth of the nuclear diameter. Thus the range of vibrational wave-lengths for which the statistical treatment is valid is almost vanishingly small! ($R \gg \lambda \gg \frac{1}{6}R$.) The inclusion of elastic compression vibrations is unlikely to result in a large reduction in level spacing for excitation energies less than 10 MeV. These may be treated in a manner similar to the surface waves (the empirically determined quantity needed in place of ζ is the bulk modulus of elasticity K ($K \propto \partial^2 E/\partial R^2$)), and their inclusion reduces $D(E)$ by a factor of from 3 to 10, depending on the values of K, r_0 and ζ adopted.

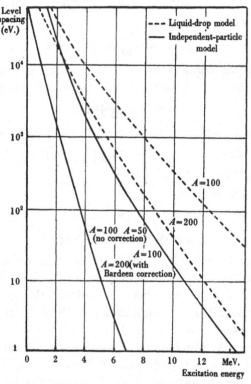

Fig. 27. Density of levels for heavy nuclei.

To summarize, then, one can say that all statistical treatments give similar variation of $D(E)$ with E, but the models are not sufficiently reliable, both as regards the approximations made and magnitude of the parameters involved, to provide accurate values for $D(E)$. A general relation such as

$$D(E) = C \exp(-\alpha E^\gamma)$$

suffices to fit all available data and models if C, and α, are adjusted empirically! (Slight variation of C and α with A may be necessary.) γ is approximately $\frac{1}{2}$ on all models. Fig. 27 shows the variation of T and $D(E)$ with E for some typical nuclei according to some models.

22. SOME EXPERIMENTAL REGULARITIES

Much, if not most, of the existing experimental information may be described as being in loosely defined 'agreement' with imprecisely formulated theory. There is no remarkable detailed agreement between theory and experiment, but there is likewise, alas, no violent disagreement. It could be concluded that the general nature of the problem of nuclear spectra is well understood, but that the details are so complex that precise theoretical predictions might, if possible at all, require the expenditure effort out of proportion to their value. It is of interest then to notice that there are some indications, albeit slight, of divergence between theory and experiment which may require more than the adjustment of some arbitrary parameters and whose elucidation may provide some substantial advance in our knowledge of nuclear structure.

22·1. Integral relations between excitation energies

Several investigators have pointed out the frequent occurrence of simple integral relations between the energies of the levels of an individual nucleus. High accuracy of measurement (and small integers!) is essential if merely accidental approximate integral relations are to be distinguished from genuine results. In this respect observations on γ-spectra following β-decay (§ 5·2) are probably the most reliable. One of the best known and most studied examples is the level structure of Mg^{24} formed in excited states as a result of the β-decay of Na^{24}. Two γ-transitions occur with energies (measured to an accuracy of at least 1 %) of 1·38 and 2·76 MeV. There seems little doubt that the γ-transitions are in cascade(138). Inelastic scattering of protons by Mg indicates levels of Mg^{24} at 1·35 ± 0·02 and 2·76 ± 0·02 MeV. These results together indicate level schemes of Mg^{24} either as in (a_1) or (a_2) (fig. 8). In either case there are three levels with energies 1·38 × k MeV., where $k = 1, 2, 3$.

Similar examples are found in the spectra of Fe^{56} (35), excited in the β-decay of Mn^{56} and Co^{56}, the levels observed having energies given by $0.425 \times k$ MeV., $k = 2, 3, 5 \pm 1\%$; in Mn^{56} (proton groups from the Mn^{55} (d, p) reaction (16), which has levels at $0.355k$ MeV., $k = 2, 5, 7, 10, 12$ (in this case these levels are the only observed levels, and the measured positions deviate from the simple relation by less than experimental error for the first three levels and by not more than 0.1 MeV. for the remainder); in Ge^{72}, with three γ-transitions having energies $0.21k$ MeV., $k = 3, 4, 10$; and in Xe^{130}, excited by β-decay of I^{130}, with γ-transitions having energies $k \times 0.107$, $k = 4, 5, 7 \pm 1\%$ MeV.

The greater level density in heavier nuclei makes it difficult to establish such integral relations, if, indeed, they exist. The explanation which immediately suggests itself for these regularities is that such levels are due to rotation of the whole nucleus, with the energy levels forming a series $\alpha J(J + 1)$, or following more complex formulae for non-spherical nuclei. Quantitatively, the rotational energies are of the correct order of magnitude, e.g. $J = 2, 3$ (expression in § 20·4) gives levels at 1.98 and 3.95 MeV. for Mg^{24}, but it would be rather surprising if the interaction between vibrational and rotational energies were so small that the nucleus could be treated as rigid to such a high degree of accuracy. It is also noteworthy that all the regularities mentioned above are for nuclei of the even Z-even N type which have zero angular momentum in the ground state and are usually particularly stable.* Clearly both more accurate and a more representative range of experimental data is needed to elucidate this problem.

22·2. Constancy of level spacing

In most reactions of the type (d, p) and (d, n) a study of the proton or neutron groups has indicated that the level spacing, far from decreasing rapidly as might be expected theoretically, is practically constant up to excitation energies of the order 5 MeV. or more. For example, the neutron groups from $F^{19}(d, n)Ne^{20}$ indicate energy differences between successive levels of $1.5, 2.7, 1.3, 1.9$,

---

* Note, however, that these nuclei, being stable, are frequently the end-products of β-disintegration and as such have been studied more extensively than other types.

1·7 and 1·1 MeV. (6). Proton groups from $Ne(d,p)$, S^{32} (d,p) and Mn^{55} (d, p) indicate level spacings for Ne^{21} of 0·31, 1·44, 1·08, 0·75 MeV. (139), for Ne^{23} of 0·99 and 0·67 MeV., for S^{33} of 1·05, 1·12, 1·05 and 1·11 MeV. (140) and for Mn^{56} of 1·01, 0·70, 0·71, 1·13 and 0·77 MeV. (16).

The large apparent level spacing in Ne^{20} is particularly interesting, since for excitation energies up to about 10 MeV. the (d,p) reaction indicates a constant level spacing of about 1·0 MeV., whereas for excitation energies of 13·5 MeV. the level spacing as deduced from proton capture reactions is certainly not greater than 0·1 MeV. and probably several times smaller.

Similar indications of constant level spacing are found in the excitation of light nuclei by inelastic scattering and the excitation of heavy nuclei by electrons and X-radiation. In none of these processes would one expect to observe all the nuclear levels; some transitions may be completely forbidden, or, what is more likely, occur with too small an intensity to be detected. What is so surprising is that the levels that are excited should be spaced at such regular energy intervals.

22·3. Density of resonance levels

The density of slow-neutron resonance levels appears, on the basis of the incomplete information available, to be greater in compound nuclei of the odd Z-odd N type than in other types. Thus in twenty-four elements ranging from Ru to Pb, the average number of observed resonance levels per isotope in the range 0–20 eV.* is 1·2 for elements with odd Z (these elements nearly all have either one isotope or two isotopes of comparable abundance), but only 0·25 for elements with even Z (these elements usually have four or five isotopes of comparable abundance of which about two-thirds have even A). This fact may be associated with the generally larger binding energy of nuclei with even Z than those with odd Z, resulting in a higher excitation energy of the compound nucleus when a slow neutron is captured by a nucleus with odd Z. A more complete comparison of level densities in nuclei of these two types would require, primarily, better data on the level densities for

---

* At energies higher than 20 eV. resonance absorption becomes too small to be reliably detected with the experimental resolutions actually used.

neutron energies greater than 20 eV., but this would necessitate a higher neutron energy resolution than has hitherto been attained.

The resonance-level density observed in α-particle disintegration and scattering with light nuclei ($Z < 20$) is usually much smaller than that indicated by other experimental data. For example [141], the reactions $\mathrm{Al}^{27}(\alpha, p)$ and $\mathrm{Al}^{27}(\alpha, n)$ both show resonance levels in P^{31} with mean excitation energy of about 13·5 MeV. and an average spacing of 0·4 MeV., whereas the resonance capture of protons in $\mathrm{Al}^{27}(p, \gamma)$ shows a level density in Si^{28} at a mean excitation energy of 12·5 MeV. of only 0·04 MeV., and similar results are observed in the reaction $\mathrm{Na}^{23}(p, \gamma)\,\mathrm{Mg}^{24}$. The level spacing observed for proton capture is also about the same as that observed for 'fast' neutron capture and scattering; for example, the total cross-section for the interaction of neutrons of energies up to 2 MeV. with Al^{27} shows a resonance-level spacing of 100 MeV. corresponding to an excitation energy of Al^{28} of about 9 MeV. [66].

The experiments with α-particles were made using weak natural radioactive sources with which only moderate energy resolution was possible. It may be that the apparent resonance levels are not due to individual levels but to large fluctuations in level density with single levels unresolved, or simply that only a few of the most strongly excited levels are observed.

22·4. Occurrence of isomeric states

The fact that many nuclei, particularly those with intermediate mass, have what is probably the *first* excited level at an energy of the order 10^4–10^5 eV. and a total angular momentum differing by 3 or 4 units from that of the ground state, seems now to be well established. These facts have already been discussed (§ 20·4), but it is noteworthy that there are other interesting features associated with the occurrence of these isomeric states, apart from their distribution amongst nuclei with particular odd-even characteristics. One of these is the occurrence of *similar* isomeric states in pairs of isotopic nuclei of odd Z (e.g. Ag^{107}, Ag^{109}; In^{113}, In^{115}) [142, 143] differing in mass by two units. There are even more remarkable features of such isotopic pairs, notably the fact that the two isotopes usually have the same total angular momentum (e.g. $\mathrm{Br}^{79,81}$, $\tfrac{3}{2}$;

$Ag^{107,109}$, $\frac{1}{2}$; $In^{113,115}$, $\frac{9}{2}$; but Rb^{85}, $\frac{3}{2}$ and Rb^{87}, $\frac{5}{2}$), and when this is the case the two isotopes have almost identical magnetic moments (e.g. Br, $\mu_{81}/\mu_{79} = 1\cdot0778$; In, $\mu_{115}/\mu_{113} = 1\cdot002$; but Rb, $\mu_{85}/\mu_{87} = 0\cdot491$). This last example clearly illustrates how knowledge and interpretation of nuclear spectra are intimately interwoven with knowledge of nuclear properties in general.

REFERENCES

(1) BETHE. *Rev. Mod. Phys.* **9**, no. 2, p. 69 (1937).
(2) BETHE and KONOPINSKI. *Phys. Rev.* **54**, 130 (1938).
(3) WEISSKOPF. *Phys. Rev.* **52**, 295 (1937).
(4) WEISSKOPF and EWING. *Phys. Rev.* **57**, 472 (1940).
(5) CHADWICK, MAY, PICKAVANCE and POWELL. *Proc. Roy. Soc.* A, **183**, 1 (1945).
(6) BONNER. *Proc. Roy. Soc.* A, **174**, 339 (1940).
(7) GIBSON and LIVESEY. *Proc. Phys. Soc.* **60**, 523 (1948).
(8) FEENBERG. *Phys. Rev.* **57**, 348 (1940).
   WEISSKOPF. *Phys. Rev.* **53**, 1018 (1938).
(9) DICKE and MARSHALL. *Phys. Rev.* **63**, 86 (1943).
(10) WILKINS. *Phys. Rev.* **60**, 365 (1941).
   WRENSHALL. *Phys. Rev.* **63**, 56 (1943).
(11) OPPENHEIMER and PHILLIPS. *Phys. Rev.* **48**, 500 (1935).
(12) BETHE. *Phys. Rev.* **53**, 39 (1938).
(13) VOLKOFF. *Phys. Rev.* **57**, 566 (1940).
(14) LIVINGSTON and BETHE. *Rev. Mod. Phys.* **9**, no. 3, p. 323 (1937).
(15) POWELL. *Proc. Roy. Soc.* A, **181**, 344 (1944).
   CHAMPION and POWELL. *Proc. Roy. Soc.* A, **183**, 64 (1945).
   LATTES, FOWLER and CUER. *Proc. Phys. Soc.* **59**, 883 (1947).
(16) MARTIN. *Phys. Rev.* **73**, 738 (1947).
(17) BOWER and BURCHAM. *Proc. Roy. Soc.* A, **173**, 379 (1939).
(18) LIVINGSTON and BETHE. *Rev. Mod. Phys.* **9**, no. 3, p. 292 (1937).
   COCKCROFT and LEWIS. *Proc. Roy. Soc.* A, **154**, 261 (1936).
(19) RISSER, LARK-HOROWITZ and SMITH. *Phys. Rev.* **57**, 355 (1940).
(20) BØGGILD. *K. danske vidensk. Selsk.* **23**, 4 (1945).
   GILBERT. *Proc. Camb. Phil. Soc.* (in the Press).
(21) BONNER and BRUBAKER. *Phys. Rev.* **50**, 781 (1936).
(22) CURRAN, DEE and STROTHERS. *Proc. Roy. Soc.* A, **174**, 546 (1940).
(23) FOWLER, LAURITSEN and LAURITSEN. *Rev. Mod. Phys.* **20**, 236 (1948).
(24) McDANIEL, VON DARDEL and WALKER. *Phys. Rev.* **72**, 985 (1947).
(25) HUDSON, HERB and PLAIN. *Phys. Rev.* **57**, 187 (1940).
(26) GOLDHABER, HILL and SZILARD. *Phys. Rev.* **56**, 47 (1939).
   COHEN. *Nature, Lond.* **161**, 475 (1948).
(27) BARNES and ARADINE. *Phys. Rev.* **56**, 80 (1939).
(28) PRESTON. *Phys. Rev.* **71**, 865 (1947).
(29) RUTHERFORD, WYNN-WILLIAMS, LEWIS and BOWDEN. *Proc. Roy. Soc.* A, **139**, 617 (1933); **145**, 235 (1934).
   ROSENBLUM, GUILLOT and PEREY. *C.R. Acad. Sci., Paris,* **204**, 175 (1937).
(30) CHANG. *Phys. Rev.* **69**, 60 (1946); **70**, 632 (1946).
(31) BOTHE. *Z. Phys.* **96**, 607 (1935); **100**, 273 (1936).
(32) ELLIS. *Proc. Roy. Soc.* A, **143**, 350 (1934).
(33) KONOPINSKI. *Rev. Mod. Phys.* **15**, no. 4, p. 209 (1943).

(34) SEABORG. *Rev. Mod. Phys.* **16**, no. 1, p. 1 (1944).

(35) ELLIOTT and DEUTSCH. *Phys. Rev.* **65**, 321 (1943).

(36) DUNWORTH. *Rev. Sci. Instrum.* **11**, 167 (1940).

(37) MITCHELL. *Rev. Mod. Phys.* **20**, 296 (1948).

(38) ELLIOTT, DEUTSCH and ROBERTS. *Phys. Rev.* **63**, 386 (1943).

(39) BRADT *et al. Helv. Phys. Acta*, **19**, 77 (1946).
MAIER-LEIBNITZ. *Z. Naturforsch.* **1**, 243 (1946).

(40) WIEDENBECK. *Phys. Rev.* **72**, 429 (1947).

(41) HAMILTON. *Phys. Rev.* **58**, 122 (1940).

(42) GOERTZEL. *Phys. Rev.* **70**, 897 (1946).

(43) BRADY and DEUTSCH. *Phys. Rev.* **72**, 870 (1947).

(44) BOTHE and GENTNER. *Naturwissenschaften*, **25**, 90 (1937).

(45) HIRZEL and WÄFFLER. *Helv. Phys. Acta*, **20**, 373 (1947).

(46) GUTH. *Phys. Rev.* **59**, 325 (1941).

(47) WIEDENBECK. *Phys. Rev.* **67**, 53, 92 (1945); **68**, 111, 237 (1945).

(48) MATTAUCH. *Z. Phys.* **117**, 246 (1941).

(49) HEITLER. *Quantum Theory of Radiation*, p. 263 (Oxford Univ. Press, 1944).

(50) BETHE. *Rev. Mod. Phys.* **9**, no. 2, p. 75 (1937).
PEIERLS. *Rep. Progr. Phys.* **7**, 87 (1940).

(51) GOLDSMITH, IBSER and FELD. *Rev. Mod. Phys.* **19**, 259 (1947).

(52) BETHE. *Rev. Mod. Phys.* **9**, 119 (1937).

(53) BETHE. *Rev. Mod. Phys.* **9**, 134 (1937).
HORNBOSTEL, GOLDSMITH and MANLEY. *Phys. Rev.* **58**, 18 (1940).

(54) GOLDSMITH and RASETTI. *Phys. Rev.* **50**, 328 (1936).
v. HALBAN and PREISWERCK. *Helv. Phys. Acta*, **9**, 318 (1936).

(55) MANLEY, GOLDSMITH and SCHWINGER. *Phys. Rev.* **55**, 39 (1939).

(56) DUNNING *et al. Phys. Rev.* **49**, 103, 199; **50**, 738 (1936).

(57) RAINWATER and HAVENS. *Phys. Rev.* **70**, 136, 154 (1947); **71**, 65, 165 (1947).

(58) MCDANIEL. *Phys. Rev.* **70**, 832 (1946).

(59) MEYERS *et al. Phys. Rev.* **69**, 666 (1946).

(60) DEMPSTER. *Phys. Rev.* **71**, 829 (1947).

(61) ZINN. *Phys. Rev.* **71**, 752 (1947).

(62) STURM. *Phys. Rev.* **71**, 757 (1947).

(63) SAWYER *et al. Phys. Rev.* **72**, 110 (1947).

(64) FESHBACH, PEASLEE and WEISSKOPF. *Phys. Rev.* **71**, 145 (1947).

(65) STAUB and TATEL. *Phys. Rev.* **58**, 820 (1940).

(66) SEAGONDOLLAR and BARSCHALL. *Phys. Rev.* **72**, 439 (1947).

(67) HERB, KERST and MCKIBBEN. *Phys. Rev.* **51**, 691 (1937).

(68) BURCHAM and DEVONS. *Proc. Roy. Soc.* A, **173**, 555 (1939).

(69) BAILEY, PHILLIPS and WILLIAMS. *Phys. Rev.* **62**, 80 (1942).
BENNETT *et al. Phys. Rev.* **59**, 781 (1941).

(70) BROSTROM, HUUS and TANGEN. *Phys. Rev.* **71**, 663 (1947).

(71) BENDER, SHOEMAKER and POWELL. *Phys. Rev.* **71**, 905 (1947).

(72) FOWLER, LAURITSEN and LAURITSEN. *Rev. Mod. Phys.* **20**, 236 (1948).

(73) BONNER and EVANS. *Phys. Rev.* **73**, 666 (1948).

(74) DEVONS and HEREWARD. *Nature, Lond.* **162**, 331 (1948).

(75) STREIB, FOWLER and LAURITSEN. *Phys. Rev.* **59**, 253 (1941).

(76) LIVINGSTON and BETHE. *Rev. Mod. Phys.* **9**, no. 3, p. 295 (1937).
(77) BENNETT *et al.* *Phys. Rev.* **59**, 781 (1941).
(78) BETHE. *Rev. Mod. Phys.* **9**, 173 (1937).
(79) ROSE. *Phys. Rev.* **57**, 958 (1940).
(80) BRUBAKER. *Phys. Rev.* **54**, 1011 (1938).
      DEVONS. *Proc. Roy. Soc.* A, **172**, 127 (1939).
(81) DEVONS. *Proc. Roy. Soc.* A, **172**, 559 (1939).
(82) WHEELER. *Phys. Rev.* **59**, 27 (1941).
(83) CREUTZ. *Phys. Rev.* **55**, 819 (1939).
(84) RUBIN. *Phys. Rev.* **72**, 1176 (1947).
(85) DEE and GILBERT. *Proc. Roy. Soc.* A, **154**, 279 (1936).
(86) HEITLER. *Quantum Theory of Radiation*, Chap. v (Oxford Univ. Press, 1945).
(87) CURRAN, DEE and PETRIZILKA. *Proc. Roy. Soc.* A, **169**, 269 (1938).
(88) LATYSHEV. *Rev. Mod. Phys.* **19**, 132 (1947).
(89) SCHARFF-GOLDHABER. *Phys. Rev.* **59**, 937 (1941).
(90) RUTHERFORD, CHADWICK and ELLIS. *Radiations from Radioactive Substances*, Chap. XII (Camb. Univ. Press, 1930).
(91) ROBERTS *et al.* *Phys. Rev.* **64**, 270 (1943).
(92) DU MOND. *Rev. Sci. Instrum.* **18**, 626 (1947).
(93) HELMHOLTZ. *Phys. Rev.* **60**, 188 (1941).
(94) ELLIS and MOTT. *Proc. Roy. Soc.* A, **139**, 374 (1933).
(95) FLAMMERSFELD. *Z. Phys.* **114**, 227 (1939).
(96) HEITLER. *Quantum Theory of Radiation*, Chap. III (Oxford Univ. Press, 1945).
(97) DANCOFF and MORRISON. *Phys. Rev.* **55**, 122 (1939).
(98) BETHE. *Rev. Mod. Phys.* **9**, 87 (1937).
(99) HULME, MOTT, OPPENHEIMER and TAYLOR. *Proc. Roy. Soc.* A, **155**, 315 (1936).
(100) HEBB and UHLENBECK. *Physica*, **7**, 605 (1938).
(101) WEISZÄCKER. *Naturwissenschaften*, **24**, 813 (1936).
(102) DANCOFF and MORRISON. *Phys. Rev.* **55**, 122 (1939).
(103) HEBB and NELSON. *Phys. Rev.* **58**, 486 (1940).
(104) JAEGER and HULME. *Proc. Roy. Soc.* A, **148**, 708 (1935).
(105) SACHS. *Phys. Rev.* **57**, 194 (1940).
(106) CRANE and HALPERN. *Phys. Rev.* **55**, 260 (1939).
(107) BETHE. *Rev. Mod. Phys.* **9**, 226 (1937).
(108) BERTHOLET. *Ann. d. Physique*, **19**, 117 (1944).
       WIEDENBECK. *Phys. Rev.* **69**, 567 (1946).
(109) BITTENCOURT and GOLDHABER. *Phys. Rev.* **70**, 780 (1946).
(110) DE' BENEDETTI and McGOWAN. *Phys. Rev.* **70**, 569 (1946).
(111) HIRTZEL, STOLL and WÄFFLER. *Helv. Phys. Acta*, **20**, 241 (1947).
(112) BETHE and BACHER. *Rev. Mod. Phys.* **8**, §§ 19–23 (1936).
(113) WIGNER. *Phys. Rev.* **51**, 106, 947 (1937).
(114) WIGNER and FEENBERG. *Rep. Progr. Phys.* **8**, 274 (1942).
       KONOPINSKI. *Rev. Mod. Phys.* **15**, 7 (1943).
       FEENBERG. *Rev. Mod. Phys.* **17**, 255 (1947).
(115) BETHE. *Elementary Nuclear Theory* (Wiley, 1947).
(116) GRAVES. *Phys. Rev.* **57**, 885 (1940).
(117) BETHE. *Rev. Mod. Phys.* **9**, 172 (1937).

(118) BETHE and BACHER. *Rev. Mod. Phys.* **8**, 149 (1936).

(119) FEENBERG and PHILLIPS. *Phys. Rev.* **51**, 597 (1937).

(120) WIGNER and FEENBERG. *Phys. Rev.* **51**, 95 (1937).

(121) BETHE and BACHER. *Rev. Mod. Phys.* **8**, § 11 (1936).

(122) WHEELER. *Phys. Rev.* **52**, 108 (1938).

(123) WHEELER. *Phys. Rev.* **59**, 256 (1941).

(124) HAFSTAD and TELLER. *Phys. Rev.* **54**, 681 (1938).

(125) DENNISON. *Phys. Rev.* **57**, 454 (1940).

(126) HORNYAK and LAURITSEN. *Rev. Mod. Phys.* **20**, 191 (1948).

(127) FOWLER and LAURITSEN. *Phys. Rev.* **59**, 253 (1941).

(128) TELLER and WHEELER. *Phys. Rev.* **53**, 778 (1938).

(129) GUGGENHEIM. *Proc. Roy. Soc.* A, **181**, 169 (1943).

(130) DENNISON. *Rev. Mod. Phys.* **3**, 280 (1931).

(131) BETHE. *Rev. Mod. Phys.* **9**, 86 (1937).

(132) FEENBERG. *Rev. Mod. Phys.* **17**, 255 (1947).

(133) BETHE. *Rev. Mod. Phys.* **9**, 81 (1937).

(134) TOLMAN. *Principles of Statistical Mechanics*, p. 391 (Oxford Univ. Press, 1938).

(135) MARGENAU. *Phys. Rev.* **59**, 627 (1941).

(136) BARDEEN and FEENBERG. *Phys. Rev.* **54**, 809 (1938).

(137) BARDEEN. *Phys. Rev.* **51**, 779 (1937).

(138) WIEDENBECK. *Phys. Rev.* **72**, 429 (1947).

(139) ELDER, MOTZ and DAVISSON. *Phys. Rev.* **71**, 917 (1947).

(140) SMITH and POLLARD. *Phys. Rev.* **59**, 942 (1941).

(141) WARING and CHANG. *Proc. Roy. Soc.* A, **157**, 652 (1936).

(142) BRADT *et al. Helv. Phys. Acta*, **20**, 153 (1947).

(143) LAWSON and CORK. *Phys. Rev.* **57**, 982 (1940).

SUBJECT INDEX

Printed in the United States
By Bookmasters